List of titles

Already published

Cell Differentiation	J.M. Ashworth
Biochemical Genetics	R.A. Woods
Functions of Biological Membranes	M. Davies
Cellular Development	D. Garrod
Brain Biochemistry	H.S. Bachelard
Immunochemistry	M.W. Steward
The Selectivity of Drugs	A. Albert
Biomechanics	R. McN. Alexander
Molecular Virology	T.H. Pennington, D.A. Ritchie
Hormone Action	A. Malkinson
Cellular Recognition	M.F. Greaves
Cytogenetics of Man and other Animals	A. McDermott
RNA Biosynthesis	R.H. Burdon
Protein Biosynthesis	A.E. Smith

In preparation

The Cell Cycle	S. Shall
Biological Energy Transduction	C. Jones
Control of Enzyme Activity	P. Cohen
Metabolic Regulation	R. Denton, C.I. Pogson
Polysaccharides	D.A. Rees
Microbial Metabolism	H. Dalton
Bacterial Taxonomy	D. Jones
Molecular Evolution	W. Fitch
A Biochemical Approach to Nutrition	R.A. Freedland
Metal Ions in Biology	P.M. Harrison, R. Hoare
Nitrogen Metabolism in Plants and Microorganisms	A.P. Sims
Cellular Immunology	D. Katz
Muscle	R.M. Simmons
Xenobiotics	D.V. Parke
Plant Cytogenetics	D.M. Moore
Human Genetics	J.H. Edwards
Population Genetics	L.M. Cook
Membrane Biogenesis	J. Haslam
Biochemical Systematics	J.H. Harborne
Biochemical Pharmacology	B.A. Callingham
Insect Biochemistry	H.H. Rees

OUTLINE STUDIES IN BIOLOGY

Editor's Foreword

The student of biological science in his final years as an undergraduate and his first years as a graduate is expected to gain some familiarity with current research at the frontiers of his discipline. New research work is published in a perplexing diversity of publications and is inevitably concerned with the minutiae of the subject. The sheer number of research journals and papers also causes confusion and difficulties of assimilation. Review articles usually presuppose a background knowledge of the field and are inevitably rather restricted in scope. There is thus a need for short but authoritative introductions to those areas of modern biological research which are either not dealt with in standard introductory textbooks or are not dealt with in sufficient detail to enable the student to go on from them to read scholarly reviews with profit. This series of books is designed to satisfy this need. The authors have been asked to produce a brief outline of their subject assuming that their readers will have read and remembered much of a standard introductory textbook of biology. This outline then sets out to provide by building on this basis, the conceptual framework within which modern research work is progressing and aims to give the reader an indication of the problems, both conceptual and practical, which must be overcome if progress is to be maintained. We hope that students will go on to read the more detailed reviews and articles to which reference is made with a greater insight and understanding of how they fit into the overall scheme of modern research effort and may thus be helped to choose where to make their own contribution to this effort. These books are guidebooks, not textbooks. Modern research pays scant regard for the academic divisions into which biological teaching and introductory textbooks must, to a certain extent, be divided. We have thus concentrated in this series on providing guides to those areas which fall between, or which involve, several different academic disciplines. It is here that the gap between the textbook and the research paper is widest and where the need for guidance is greatest. In so doing we hope to have extended or supplemented but not supplanted main texts, and to have given students assitance in seeing how modern biological research is progressing, while at the same time providing a foundation for self help in the achievement of successful examination results.

J.M. Ashworth, Professor of Biology, University of Essex.

Protein Biosynthesis

Alan Smith
Department of Molecular Virology
Imperial Cancer Research Fund
Lincolns Inn Fields, London

LONDON
CHAPMAN AND HALL

A Halsted Press Book
JOHN WILEY & SONS, INC., NEW YORK

First published in 1976
by Chapman and Hall Ltd
11 New Fetter Lane, London EC4P 4EE
© *1976 Alan Smith*
Printed in Great Britain by
William Clowes & Sons Ltd.,
London, Colchester and Beccles

ISBN 0 412 13460 8

This title is available in both hardbond and paperback editions. The paperback edition is sold subject to the condition that it shall not, by way of trade or otherwise, be lent, re-sold, hired out, or otherwise circulated without the publisher's prior consent in any form of binding or cover other than that in which it is published and without a similar condition including this condition being imposed on the subsequent purchaser.

All rights reserved. No part of this book may be reprinted, or reproduced or utilized in any form or by any electronic, mechanical or other means, now known or hereafter invented, including photocopying and recording, or in any information storage and retrieval system, without permission in writing from the Publisher.

Distributed in the U.S.A.
by Halsted Press, a Division
of John Wiley & Sons, Inc., New York

Library of Congress Cataloging in Publication Data
Smith, Alan E
    Protein biosynthesis.

    (Outline studies in biology series)
    "A Halsted Press book".
    Bibliography: P.
    1. Protein biosynthesis. I. Title.
QP551.S64   1976   574.1'9296   75-41516
ISBN 0-470-80263-4

Contents

1	Introduction	7
1.1	The problem	7
1.2	Overall steps in protein biosynthesis	7
2	The molecules involved in protein biosynthesis	9
2.1	Messenger RNA	9
	2.1.1 Discovery and isolation	9
	2.1.2 Assay of mRNA	10
	2.1.3 Structure	13
2.2	Ribosomes	17
	2.2.1 Introduction	17
	2.2.2 Partial reactions occurring on ribosomes	19
	2.2.3 Structure and function	22
	2.2.4 Biosynthesis	28
2.3	Transfer RNA	29
	2.3.1 Discovery and properties	29
	2.3.2 Reactions of tRNA	30
	2.3.3 Purification and primary sequence of tRNA	30
	2.3.4 Tertiary structure of tRNA	33
	2.3.5 Structure and function of tRNA	34
	2.3.6 Mutant tRNAs	36
2.4	Initiator tRNA	37
2.5	Amino-acyl-tRNA synthetases	38
2.6	Elongation factors	41
	2.6.1 Elongation factors T and I	41
	2.6.2 Elongation factors G and II	42
2.7	Peptidyl transferase	43
2.8	Initiation factors	43
2.9	Termination factors	44
	References	44

3	The mechanism of protein biosynthesis and its control	46
3.1	Introduction	46
3.2	mRNA metabolism	46
3.3	Initiation complex formation	47
	3.3.1 Binding of initiator tRNA	47
	3.3.2 Binding of messenger RNA	50
3.4	Elongation	56
3.5	Termination of protein biosynthesis and post-translational modification	59
3.6	RNA phage protein synthesis	61
	References	63
	Index	64

1 Introduction

1.1 The problem

The discovery that the genetic material of living organisms is DNA, and the later demonstration that the DNA molecule is a double helix were both great milestones in twentieth century science, and formed the foundation of the new discipline of molecular biology. But even after these momentous discoveries, the detailed mechanism by which such genetic material could be expressed as the structural and catalytic proteins which play so important a role in the functioning of all living cells was still not obvious. It was only after the rigorous demonstration that a given protein is composed of a unique linear sequence of amino acids that the possibility of a relationship between the base sequence of DNA and the sequence of amino acids in a protein was realized. Once this conceptual breakthrough had been made, the complex task of unravelling the many steps in protein biosynthesis could begin in the laboratory. This task has continued apace during the past twenty years, so that today we have a fairly clear overall picture of the process. This elucidation of the many steps and intricacies of protein biosynthesis, which is now known to involve well in excess of a hundred different molecular species, must rank as one of the major and most exciting achievements of present-day molecular biology.

The purpose of this monograph is to outline in some detail our present knowledge of protein biosynthesis, describing the molecules and the basic mechanisms involved, and also the possible control processes operating to adjust protein synthesis to the needs of the cells and organism. It will be assumed that the reader has some knowledge of molecular biology in general and protein biosynthesis in particular, but by way of introduction each of the major molecules and stages of the process will be described in simple terms, and in subsequent chapters each will be discussed again in greater depth.

1.2 Overall steps in protein biosynthesis

The information encoded in the two complementary strands of the DNA of any structural gene is transcribed by an enzyme called DNA-dependent RNA polymerase. It makes a single-stranded RNA copy, complementary to one of the strands, which is called *messenger RNA* (mRNA). This attaches to a subcellular organelle called a *ribosome* which is composed of two subunits and functions as a black box upon which the mRNA is *translated*. The term translation encompasses all those steps by which the genetic content of the mRNA contained in the linear sequence of ribonucleotides is converted into a linear sequence of amino acids which may have enzymatic or other biological properties. The language of translation which dictates that a particular sequence of nucleotides (a *codon*) leads to the insertion of one particular amino acid is called the *genetic code*. A special class of adaptor molecules which are both able to read the codons of the genetic code and to carry the appropriate amino acid to the ribosome for polymerization are composed of RNA and termed *transfer RNA* (tRNA).

The initiation of protein biosynthesis is the process by which the mRNA first attaches to a ribosome and is prepared for translation. Initiation involves a series of critical reactions, requiring a unique *initiator tRNA* and catalysed by several proteins, termed *initiation factors*, which ensure the correct alignment or *phasing* of the mRNA upon the ribosome. The formation of the *initiation complex* containing all the components mentioned above is probably one of the rate-limiting steps in the biosynthesis of a protein and it is the point at which many control elements operate.

After the successful formation of the 'initiation complex' the next amino acid in the polypeptide chain is carried to the ribosome attached to a tRNA. The binding of the appropriate *amino-acyl tRNA* is directed by the codon and is catalysed by an *elongation factor*. Peptide bond formation is then catalysed by a ribosomal protein *peptidyl-transferase* and results in the formation of a covalent bond between the amino group of the incoming amino acyl-tRNA and the carboxyl group of the amino acid immediately preceding it in the growing polypeptide. In this way the mRNA is read 5' to 3' and the polypeptide chain *elongated* from the NH_2 to COOH ends. The ribosome is then *translocated* along the mRNA by a second elongation factor such that a new codon is exposed and the cyclic elongation process then continues.

Once elongation has proceeded sufficiently and the ribosome passed some distance along the mRNA, the ribosome-binding site becomes free again and a second ribosome can attach. In this way several ribosomes may attach to a single mRNA molecule and the resulting structure is called a *polysome*. When a ribosome passes to the end of the coding portion of an mRNA it reaches a codon signalling *terminate* protein synthesis and in a reaction involving several *termination factors* the polypeptide is released. The nascent polypeptide chain probably begins to fold into its native secondary and tertiary structure whilst it is being synthesized and is still bound to the ribosome.

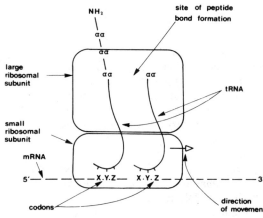

Fig. 1.1 General model of the ribosome.

Nevertheless, as released the polypeptide is not necessarily in its final form. In addition to more folding of the nascent chain and the possible formation of dilsuphide bridges, several other *post-translational modifications* may also occur.

The mechanism of protein biosynthesis is remarkably similar in different organisms. The major differences appear on moving from the prokaryotes to the eukaryotes and are mainly a consequence of the presence of the nuclear membrane in higher cells and the longer half-life of such cells. Other structural differences in the size of ribosomes and structure of the initiator tRNA, mRNA etc. are rather trivial. The genetic code itself is probably completely universal.

2 The molecules involved in protein biosynthesis

In this chapter, each of the molecules, classes of molecule or molecular components involved in protein biosynthesis will be described. Their role in protein biosynthesis, methods of isolation and results of biochemical and biophysical analysis will be discussed.

2.1 Messenger RNA

2.1.1 Discovery and isolation

Early experiments, in which rats were injected with labelled amino acids and the incorporation of radioactivity into polypeptides measured after various chase periods, indicated that amino acids were first assembled into protein in the microsomal fraction of cells, probably on ribosomes attached to membranes. The triplet nature of the genetic code was also deduced at an early date by genetic analysis, but the way in which the genetic information was transferred from DNA to the ribosome remained unclear until the concept of messenger RNA (mRNA) was crystallized in the *lac* operon theory of Jacob and Monod [1].

With this theory in mind, it was quickly established that a new species of non-ribosomal RNA was associated with *E. coli* ribosomes shortly after infection with bacteriophage T4 [2] and this added experimental support to the mRNA concept. However, it proved extremely difficult to isolate mRNA in any quantity and study its biochemical properties. At about the same time (1961), however, further support for a role for RNA in protein synthesis came with the discovery [3] that synthetic polyribonucleotides could act *in vitro* as messenger and direct the polymerization of amino acids into polypeptides. This finding was quickly exploited to elucidate the exact nature of the genetic code (see monograph by Dr. Woods for detailed discussion and Section 2.2). The visualization of polysomes in the electron microscope with a thin strand of material joining adjacent ribosomes, added further weight to the arguments in favour of the existence of a messenger molecule (Fig. 2.1).

Nevertheless, the isolation of cellular mRNA still remained a grave experimental problem, first because it constitutes only a small fraction of cellular RNA and secondly because it is very susceptible to attack by ribonucleases.

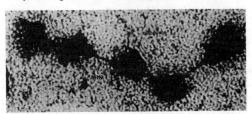

Fig. 2.1 Electron micrograph of a rabbit reticulocyte polysome.

For a long period of time putative mRNA was usually characterized as that heterogeneous population of rapidly labelled, polysome associated RNA, that could be dissociated from ribosomes by treatment with EDTA, and which had a base composition similar to that of bulk

DNA rather than ribosomes (which have a relatively high content of the ribonucleotides guanosine (G) and cytosine (C)).

The isolation of large amounts of RNA with messenger-like properties was first achieved using bacteriophage and viruses. Several coliphage (eg. f_2, R_{17}, MS_2, Q_β) contain no DNA, and their genome appears instead to consist of single stranded RNA (ssRNA), surrounded by a simple protein coat. Similar, very simple animal viruses also exist, the best known being the picornaviruses which include polio, and foot and mouth disease viruses. It was soon realized that such phage and viruses might represent one of the simplest classes of self-replicating organisms whose RNA could function both as the store of genetic information and the mRNA itself. Since the ssRNA constitutes 30% of the mass of particles, the isolation of large amounts of RNA from such sources is not difficult, and its ability to direct the synthesis of the appropriate polypeptides in a cell-free system can then be tested.

The isolation of bacterial mRNA even today presents formidable difficulties because of its inherent instability *in vivo* but one method which is being increasingly used is the enzymatic synthesis of mRNA *in vitro*. The methodology of RNA synthesis is beyond the scope of this monograph (and is discussed in Dr Burdon's book in this series), but using appropriate enzymes and cofactors the accurate and large scale production of mRNA can be achieved *in vitro* providing that a suitable DNA template is available. Early experiments used phage DNA such as T4 but as methods to amplify and purify specific bacterial genes become available, more sophiscated DNA templates are being used, such as purified *lac* operon DNA.

In the past, animal cell mRNA has been isolated from polysomes by disrupting them with a chelating agent such as EDTA [4] and separating the resulting ribosomal subunits and messenger ribonucleoprotein complex (mRNP) on a sucrose gradient. The protein can be removed from the mRNP by treatment with detergent. Many of the more complex animal viruses contain within their virion enzymes which are able to synthesize mRNA. For example, reovirus which has a fragmented double stranded RNA (dsRNA) genome contains an enzyme able to use the dsRNA as template for the synthesis of ssRNA *in vitro*. This and similar viruses have been exploited to make large amounts of eukaryotic mRNA *in vitro*. However, much the easiest and most novel method of isolating eukaryotic cell mRNA makes use of the recent discovery that a sequence of polyA is present in almost all mRNA species. Such polyA can hybridise to polyU impregnated on glass fibre filters, to polyU attached to sepharose or to oligo dT attached to cellulose, whereas other RNA species are not so bound [5]. The polyA containing RNA can be eluted by reducing the salt concentration or by using formamide, and its properties as an mRNA examined. Many eukaryotic mRNAs have been isolated in the last two or three years by utilizing the polyA tail; these include the messengers for globin, immunoglobins, myosin, ovalbumin and several viral proteins (for review see [6].)

2.1.2 Assay of mRNA

The ultimate assay of any mRNA preparation is its ability to direct the synthesis of the protein for which it codes. This can be achieved in many cases by adding the putative mRNA to an appropriate cell-free system containing radioactively labelled amino acids, and after a suitable period of incubation analysing the labelled polypeptides that have been synthesized.

Cell-free systems from bacteria are usually made from cells harvested during the exponential growth phase. These are washed and then ground with alumina in an ice-cold mortar and pestle with buffer containing mono and divalent cations (usually NH_4^+, always Mg^{++}), a sulphydryl protecting reagent and DNAse. Once the cells are disrupted, the mixture is centrifuged at 30 000 g for 10 minutes to sediment the alumina, cell wall and debris and the supernatant (which is referred to as an S30) is removed. The S30 is next pre-incubated at 37°C under the appropriate ionic conditions in the presence of unlabelled amino acids, ATP, GTP and an energy generating system. During the pre-incubation the endogenous mRNA in the S30 is translated and destroyed. After dialysis to remove the amino acids and other low molecule weight materials, the S30 is stored in small portions at low temperature (liquid nitrogen) [3].

Similar pre-incubated S30 preparations can be made from many eukaryotic cells. One that has been extensively utilized is prepared from Krebs II cells grown as an ascitic tumour in mice [7]. A very useful cell-free system which has low levels of endogenous mRNA activity and is prepared from commercial wheat germ has also been described [8]. A slightly different cell-free system is that prepared from rabbit reticulocytes. Rabbits are made anaemic and the reticulocytes prepared and washed. The packed reticulocytes, which have a very fragile membrane, are then lysed by osmotic shock and the cell stroma removed by centrifugation. The resulting supernatant which is referred to as a lysate is extremely efficient in the cell-free synthesis of globin, the initial rate of synthesis being similar to that in whole cells. Furthermore exogenous mRNAs are also efficiently translated when added to the lysate.

Each of the cell-free extracts mentioned above contain the ribosomes, enzymes, tRNA molecules etc. required for polypeptide synthesis *in vitro*. In addition to the high molecular weight components contained in the extracts amino acids, ATP, GTP and an energy-generating source are required (Fig. 2.2). The concentration of cations is also critical; synthesis is extremely sensitive to changes in Mg^{++} concentration (Fig. 2.3) and to a lesser extent to K^+ or NH_4^+ concentration. Assuming these conditions are satisfied, addition of small amounts of mRNA (1 μg) leads to the incorporation of large amounts

Cellular extract (from bacteria, animal cells, wheat germ etc.)	5 mg/ml
Tris buffer	20 mM
$MgCl_2$	1–5 mM
KCl	60–120 mM
ATP	1 mM
GTP	0.1 mM
creatine phosphate	5 mM
creatine phosphokinase	5 μg/ml
β mercaptoethanol	5 mM
amino acids (except radioactive species)	50 μM
radioactive amino acid	10–200 μCi/ml
mRNA	20–200 μg/ml

{ energy source } for creatine phosphate and creatine phosphokinase

Total volume 25 μl–1 ml
Incubation at 20–37°C
For 10–180 min

Fig. 2.2 Requirements for protein synthesis *in vitro*.

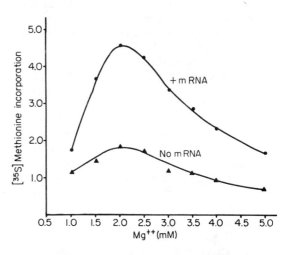

Fig. 2.3 Protein synthesis *in vitro* varies with Mg^{++} concentration (redrawn from [9]).

11

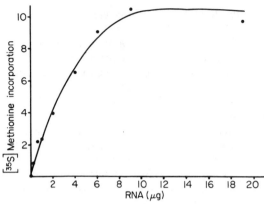

Fig. 2.4 Protein synthesis *in vitro* in response to added viral RNA (redrawn from [9]).

of labelled amino acid into polypeptide (Fig. 2.4) [9].

A somewhat different assay for mRNA activity has been described by Gurdon and his colleagues. Growing oocytes or activated eggs of *Xenopus laevis* are micro-injected with mRNA and incubated with salts and radioactive precursor. Analysis of the proteins made in the injected cells shows that protein coded for by the injected mRNA is synthesized *in ovo* for a period of several days. The oocyte system is favourable because it requires very small amounts of mRNA, the efficiency of translation is high and translation continues for very long periods [10].

The choice of the particular system to translate a given mRNA is not necessarily obvious. Homologous systems (i.e. a mouse extract to translate a mouse mRNA) are preferable in that artefacts of cell-free translation are likely to be minimized, but homologous systems are by no means essential; rabbit globin mRNA is translated quite efficiently in extracts of wheat germ! However, although such an assay can be extremely useful to identify an mRNA it does not necessarily indicate anything about the mechanism and requirements of translation

in vivo. In general cell-free protein synthesis is extremely inefficient when compared with the process in whole cells and this must always be borne in mind when assessing the physiological significance of data obtained *in vitro*.

The incorporation of labelled amino acid into hot TCA insoluble material (polypeptide), does not indicate that mRNA is being translated accurately *in vitro*, and therefore further analysis is necessary. The commonest method used to analyse the polypeptides made in a cell-free extract is polyacrylamide gel electrophoresis. Ideally, the whole cell-free reaction is reduced and briefly boiled in a solution containing SDS. All polypeptides are thereby denatured and saturated with SDS; they subsequently migrate in the gel depending on their molecular weight. After electrophoresis the gel is stained, destained, dried and auto-

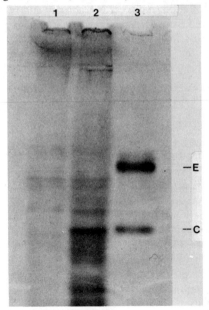

Fig. 2.5 Polyacrylamide gel of protein made *in vitro* in response to Semliki Forest Virus RNA. (1) no mRNA added (2) plus RNA (3) virus proteins; E envelope C capsid (from [9]).

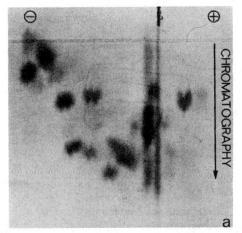
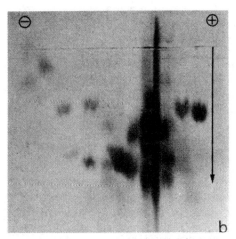

Fig. 2.6 Tryptic peptide fingerprint of (a) mouse globin (stained with ninhydrin) and (b) globin made *in vitro* using ^{14}C amino acids (autoradiograph) taken from [7].

radiographed. Comparison of the radioactively labelled polypeptides in the cell-free product with those present in similar incubations without added mRNA and with appropriate marker polypeptides is used to demonstrate the mRNA-dependent synthesis of the protein in question (Fig. 2.5).

Similarity in the molecular weight of two proteins, however, does not prove their identity and other methods must also be used. Such methods include column chromatography and immunological precipitation of protein made *in vitro*. The most rigorous method, however, is by analysis of the peptides released after digestion with a specific protease. For example trypsin cleaves polypeptides to the C terminal side of arginine and lysine residues. The cell-free polypeptide is therefore digested with trypsin, preferably after previous partial purification by one of the methods mentioned above, and the resulting peptides 'fingerprinted' by separation of the peptides in two dimensions using appropriate combinations of chromatography and/or electrophoresis (Fig. 2.6). Only after several of the techniques described have been used and proved positive can the accurate cell-free synthesis of a given protein be claimed.

2.1.3 Structure

An estimate of the size of purified mRNA can be obtained by sucrose gradient centrifugation or more accurately by polyacrylamide gel electrophoresis under denaturing conditions (for example in the presence of formamide). The size is calculated using a relationship between the distance moved by any particular molecule and the logarithm of its molecular weight. Almost invariably the molecular weight obtained for a given mRNA is greater than that expected for a molecule containing only coding information for the protein in question. For example, the number of amino acids in the α and β chains of rabbit globin are 141 and 146 respectively; allowing three nucleotides per amino acid a molecular weight of about 135 000 daltons would be predicted for the mRNA. In fact the best estimates to date suggest the true molecular weights of the mRNAs are 202 000 and 220 000 daltons [11]. Even allowing for the presence of 50 to 200 nucleotides present in polyA, and the possible loss of part of a nascent polypeptide by post-translational cleavage, all mRNAs appear, like globin, to have non-coding portions. These presumably are required to fold the mRNA, to form a site to which ribosomes can attach and to bind other transcriptional or translational control elements.

In an attempt to learn more about such features of the primary and secondary structure of mRNA, several have been analysed by nucleotide sequencing methods. Classical sequencing methods using optical density amounts of RNA are quite impractical for analysis of mRNA, but the Sanger techniques can be used. One limitation on the latter method however is that the RNA must be labelled at high specific activity with radioactive phosphorus (^{32}P) [12].

Bacteriophage mRNAs can be labelled in this way and spectacular progress has been made over the last few years in the sequence analysis of these molecules. The great length of the ssRNA phage mRNA presents enormous technical problems and in the initial sequence analysis specific portions of mRNA have been isolated and examined. Of particular interest is that region of the mRNA around the ribosome binding site. This can be isolated by first forming an initiation complex containing ribosomes, mRNA, initiator tRNA and initiation factors (but not amino acyl-tRNA or elongation factors) and then enzymatically digesting away all the mRNA not attached to and therefore

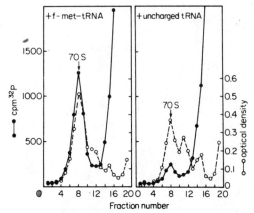

Fig. 2.7 Isolation of ribosome binding sites of R_{17} RNA. Initiation complex formed in the presence of (1) charged tRNA and (2) uncharged tRNA, digested with RNAse and separated on sucrose gradient (from [13]).

protected by the ribosome (Fig. 2.7). Several such bacterial ribosome binding sites have been isolated and sequenced (Fig. 2.8). The sequences are known to be meaningful because they all contain an AUG initiation codon followed by a nucleotide sequence which codes for the amino acids found at the N-terminus of the proteins in question. In many cases the initiating AUG codon appears at the end of a 'hairpin' loop (Fig. 2.9) where presumably it is easily recognized by ribosomes, initiation factors and initiator tRNA, and is thereby able to set the reading frame or phase of translation. Several nucleotide sequences appear common to many

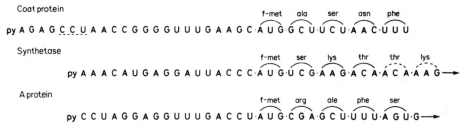

Fig. 2.8 Sequences of R_{17} ribosome binding sites (from [13]).

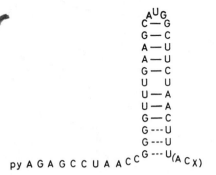

Fig. 2.9 Possible secondary structures of R17 coat protein ribosome binding site (from [13]).

of the binding sites; these may be involved as specific recognition sites. In particular the 3′ terminus of 16S ribosomal RNA (ACCUCC-UAA$_{OH}$) is complementary to one of these common sequences (GGAGGU) [14] and this region may be directly involved in ribosome binding (see Section 3.3.2).

ssRNA phage messenger codes for three proteins; the replicase, the coat protein and the maturation protein. The ribosome binding site of each of these genes has been sequenced. These then provide markers to order the three genes along the mRNA. The RNA is first cleaved specifically into two halves using specific RNAses, and the fragments identified as coming from either the 5′ or 3′-end of the mRNA by assaying for pppXp (which can only come from the 5′ terminus) and subsequently isolating the ribosome binding sites contained in both the fragments. Such analysis suggested the gene order along the mRNA is maturation: coat: replicase (Fig. 2.10).

Other methods of isolating specific fragments for sequence analysis include the identification of terminal oligonucleotides of the phage RNA, and the isolation by polyacrylamide gel electrophoresis of segments of the molecule partially resistant to ribonuclease attack (presumably they are protected by their secondary struc-

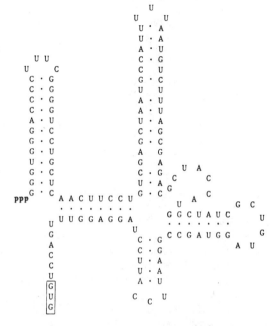

Fig. 2.11 Nucleotide sequence at the 5′ terminus of MS2 RNA (from [16]).

ture). In this way much of the nucleotide sequence of several phage mRNAs have been determined. For example the initiation codon of the maturation protein gene is located 130 nucleotides from the 5′-end of MS2 RNA, thus confirming the presence of untranslated regions (Fig. 2.11). The complete nucleotide sequence of the MS2 coat protein gene has also been

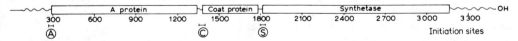

Fig. 2.10 Biochemical map of the R17 genome (from [15]).

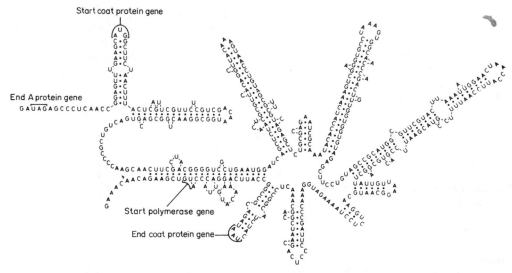

Fig. 2.12 MS2 coat protein gene (from [18]).

established. This can be compared with the amino acid sequence of the coat and this data can be used to compare the genetic code as used in natural mRNA with that obtained with synthetic polymers and other genetic methods [17].

The coat protein gene can be folded into a variety of secondary structures. Analysis of the stability of the possible structures and consideration of the susceptibility of the various parts of the molecule to digestion with RNAse lead Fiers and co-workers to propose the model shown in Fig. 2.12 for the MS2 coat protein gene. The postulated structure shows extensive 'flower-like' folding, as well as details of the initiation, termination and inter-cistronic sites of the gene. The initiation site of the polymerase gene on the mRNA is partially covered by secondary structure with the coat protein gene. This finding explains the earlier observation that ribosomes do not attach to the polymerase initiation site until some time after initiation at the coat protein gene. Presumably until a ribosome has moved down the coat protein gene

and thereby disrupted its secondary structure the polymerase ribosome binding site is inaccessible.

Other ssRNA phage messengers have been analysed in this way but all these may be considered atypical of cellular mRNAs because in addition to their messenger function they have to be packaged in the phage particle.

Analysis of other mRNAs, which cannot be extracted in sufficient quantity from whole cells or which, in the case of animal cells, cannot be labelled to high specific activity, requires other methods. One approach is to use enzymatic synthesis of the mRNA *in vitro* using purified DNA or DNA fragments, highly purified DNA dependent RNA polymerase and nucleotide triphosphates labelled in the α phosphate position. Alternatively, radioactive ^{32}P or ^{125}I can be introduced either enzymically or chemically into specific positions of unlabelled mRNA and the resulting material analysed by modifications of the Sanger technique [12].

mRNA isolated by gentle methods from animal cells contains associated protein. The function of the protein portion of mRNP is not known in most cases, but it does not appear to be essential for cell-free protein synthesis, nor does the protein appear to be related to that attached to RNA present in the nucleus [4].

The function of the polyA present in eukaryotic mRNA also remains unknown [19]. The polyA is normally between 50–200 residues long and is attached to the 3' terminus of mRNA. It is present in most eukaryotic mRNA including mitochondrial and viral mRNA but not in histone mRNA, reovirus or cardiovirus RNA. It is also present in heterogeneous nuclear RNA (HnRNA) which is the probable nuclear precursor to cytoplasmic mRNA. Treatment of cells with cordycepin (3 deoxyadenosine) inhibits the addition of polyA to nuclear RNA and also the transport of mRNA to the cytoplasm. This therefore suggests polyA is involved in mRNA transport from the nucleus; however the presence of this appendage in the mRNAs of viruses which replicate solely in the cytoplasm (eg. poxviruses and enteroviruses) weakens this hypothesis. Alternatively, polyA might be involved in binding mRNA to membranes, the binding site of proteins attached to mRNA or a ticketing device for ageing mRNA. Evidence has been presented showing that as mRNA becomes older so its polyA tail becomes shorter [20]; and although mRNA from which the polyA has been enzymatically removed is translated in cell-free extracts, it is very much less stable when injected into *Xenopus* oocytes.

Very recently another surprising structural feature of eukaryotic mRNAs has been described, that is the presence of modified nucleotides at the 5'-end. Thus several mRNAs from different mammalian cells and viruses begin with an unusual sequence of modified nucleotides m^7G pppNmpNp .... where the m^7G ppp Nmp is present in a 5' to 5' linkage [21]. It is not yet known what role such methylated bases play but viral mRNA lacking such sequences are less active in cell-free protein synthesis, they could therefore be involved in ribosome binding of mRNA. Alternatively they may be involved during synthesis of mRNA, particularly in some viruses which contain polymerases requiring the methyl donor, S-adenosyl methionine, for active transcription *in vitro*. It is even possible that methylated bases can act as recognition sites for cleavage

Fig. 2.13 General model of eukaryotic mRNA. (from [21]). (m^7G and m^6A are guanosine and adenosine with methyl groups attached at positions 7 and 6 respectively on the purine rings, and Nm is a nucleoside with a methyl group attached to the sugar).

enzymes responsible for the conversion of high molecular weight nuclear RNA precursors (HnRNA) into cytoplasmic RNA molecules. At least two other modified bases are also present elsewhere in mammalian mRNA; their function is also unknown. The evidence so far agrees with the general model for eukaryotic mRNA shown in Fig. 2.13.

2.2 Ribosomes

2.2.1 Introduction

Ribosomes played a pioneering role in the birth of cell biology and molecular biology in that these organelles were amongst the first subcellular structures to be observed during the early development of the electron microscope, and the microsome fraction containing membranes with many attached ribosomes was one of the first preparations to be made using the then newly-developed technique of differential centrifugation. These early experiments were followed by a period of relatively little growth in our understanding of how ribosomes actually work, not because of lack of interest, but more a lack of biochemical technology. Only during the last ten years can it be said that real progress has been made in this field and soon it should be possible to stop referring to ribosomes as black-boxes, as has been so common in the past.

As mentioned earlier the ribosome was suspected as the site of protein biosynthesis on the basis of pulse-chase experiments with rat liver, but most of the definite work on structure/function relationships in ribosomes has been done using material isolated from bacteria, usually *E.coli*. This results from two main considerations; first, it was realized from an early date that protein biosynthesis is a complex multistep reaction and could only be studied in cell-free systems preferably as a series of partial reactions and secondly bacterial extracts appeared to give much more active cell-free systems and these were relatively easy to prepare in a pure form [3]. Nowadays this consideration is becoming less true and highly active eukaryotic systems are available [7]. It seems that most of the reactions taking place on the ribosome are remarkably similar in both eukaryotic and prokaryotic cells. Although in many cases the details of each particular enzyme differ, the major difference between bacteria and higher cells is the size of the ribosome. In this chapter therefore reference will be made almost exclusively to bacterial ribosomes, the inference being that unless specifically mentioned the eukaryotic equivalent is similar although at this stage probably less well characterised.

Ribosomes are normally prepared by breaking open cells in buffered salt solutions containing Mg^{++}. After removal of cell debris and large subcellular organelles, the post-mitochondrial supernatant is centrifuged at 100 000 x g for 2 hours. This yields a crude pellet of ribosomal material from which the particular class of ribosomes required may be purified. For

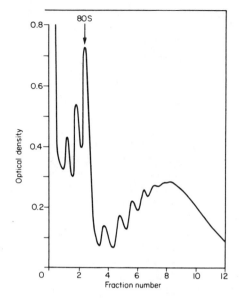

Fig. 2.14 Separation of subunits, ribosomes and polysomes on sucrose gradient (from [22]).

example, treatment with mild detergent will remove eukaryotic ribosomes from membranes and the ribosomes can then be separated from the solubilized membranes by a second centrifugation through a layer of buffered sucrose solution. Different sizes of ribosome particles and polysomes present in a crude mixture can be separated by centrifugation on appropriate sucrose density gradients (Fig. 2.14).

A further important property of ribosomes is that under certain ionic conditions *in vitro* they dissociate into sub-ribosomal particles (subunits) [23]. Thus bacterial ribosomes which normally sediment at about 70S dissociate into two subunits of 50S and 30S respectively. A certain proportion of subunits are also present in normal cells. The reason for ribosomes being composed of two subunits is not clear. Subunits play an important role in the initiation of protein synthesis, but it is possible that they also provide a mechanism to produce the relative movement within the ribosomal structure that is necessary for translocation of mRNA. Even after successively washing ribosomes through sucrose many proteins remain stuck to the particles, some of these can be removed by washing with higher concentrations of salt (e.g. 1 or 2M NH$_4$Cl). The wash fraction contains several proteins that are necessary for protein synthesis but only those proteins that remain tightly bound are considered true ribosomal proteins. Partially purified ribosomes consist of roughly half RNA and half protein, the proportions varying with different organisms. In general there is more protein associated with ribosomes from higher cells.

2.2.2 Partial reactions occurring on ribosomes

Most research into the ribosome today attempts to explain the known biosynthetic properties of the ribosome in terms of the properties of the RNA and protein of which the ribosome is composed. This is best approached by studying separately each individual step of protein biosynthesis and relating each reaction to the properties of the micro-environment of the particular site on the ribosome in which the reaction occurs.

Many of the partial reactions are considered in terms of a theoretical model for the ribosome (Fig. 2.15). For peptide bond formation to take place between two adjacent amino-acyl tRNA molecules the ribosome must have at least two sites to which tRNA can bind. Since as the peptide chain elongates, one of the tRNA molecules has the nascent peptide chain attached to it, the site to which it is bound prior to peptide bond formation is called the peptidyl-tRNA site (or P-site); the other site to which incoming amino acyl-tRNA first attaches is called the amino acyl-tRNA site (or A-site). Thus any

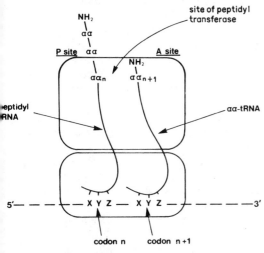

Fig. 2.15 Schematic diagram of a ribosome showing the A and P sites.

Binding initiator tRNA
Binding mRNA (synthetic/natural)
Binding amino acyl-tRNA
Peptide bond formation
Termination reaction
Association/dissociation subunits

Fig. 2.16 Partial reactions of ribosomes.

reaction to be studied can be related to the A- or the P-site on the ribosome and to the large or small ribosomal subunit. In much the same way all the reactions are broken down and related to particular areas of the ribosome with the eventual aim of relating each catalytic step to an individual ribosomal protein. Such an approach has recently had spectacular successes, but it must be remembered that a conformational change in one ribosomal protein may affect profoundly properties of proteins in distant parts of the structure. Ribosomal reactions therefore need not necessarily all involve direct molecular interactions and results must always be interpreted with this in mind.

The partial reactions occurring on ribosomes are listed in Fig. 2.16. The binding of mRNA to ribosomes can be studied using synthetic or natural messenger and measuring the binding either by sucrose density gradient analysis or passing the complex through cellulose acetate filters (Millipore filters). The particle to which the mRNA has attached will move with its characteristic mobility on sucrose gradients and can be detected by its optical density; the messenger on the other hand can be detected by radioactive labelling (Fig. 2.7). The results of such analysis must be treated with great caution because although binding may occur *in vitro* there is little guarantee that such binding is physiological unless independent checks are made. The best such check is a nucleotide fingerprint of the ribosome-protected region of the mRNA as described in Section 2.1.3.

mRNA appears initially to bind to the smaller ribosomal subunit and such binding requires initiation factors and Mg^{++} ions. The factors are required only for binding natural mRNA, and have little effect with the non-physiological synthetic polymers such as polyuridylic acid (polyU).

The binding of initiator tRNA and amino acyl-tRNA ($\alpha\alpha$-tRNA) to ribosomes can also be studied as a partial reaction *in vitro*. Again the complexes are detected on sucrose gradients or on Millipore filters, the $\alpha\alpha$-tRNA usually being detected by radioactive label in the amino acid. At high concentrations of Mg^{++} ion binding occurs non-enzymically, but at lower concentration either initiation factors or elongation factors and GTP are required to catalyse the reaction. Binding is codon directed and results in $\alpha\alpha$-tRNA being attached in the A-site of a 70S ribosome. Initiator tRNA on the other hand binds first to the smaller subunit; this does not have A- and P-sites as defined above, but once initiation is complete and the 70S ribosome assembled the initiator-tRNA is present in the P-site (see Section 3.3.1).

$\alpha\alpha$-tRNA binding is normally dependent on codons present in the messenger which is added. Leder and Nirenburg [24], however, showed that trinucleotides (triplets) could replace mRNA and direct efficient binding. This triplet-binding assay was used to confirm the genetic code by synthesizing each of the possible 64 triplets and testing which $\alpha\alpha$-tRNA was bound with each triplet (Fig. 2.17). It is also useful in assaying the enzymatic requirements for triplet-dependent $\alpha\alpha$-tRNA binding, and for initiation and termination assays. It can be used with either 70S or 30S ribosomal subunits.

Peptide bond formation is the basic synthetic step in protein biosynthesis and takes place on the ribosome when two tRNAs, one carrying a nascent peptide and the other an amino acid, lie adjacent to one another in the two ribosomal sites. The antibiotic puromycin is very useful in assaying for peptide bond formation *in vitro* because identification of the reaction products in assays using peptidyl-tRNA is rather tedious. Puromycin avoids such problems because it resembles the terminal phenylalanyl-adenosine portion of Phe-tRNA and it is therefore able to act as a substrate for the enzyme, peptidyl-transferase, by mimicking the incoming $\alpha\alpha$-tRNA (Fig. 2.18). This results in the formation of a peptidyl- or amino acyl-puromycin which is

	Second letter				
First letter	U	C	A	G	Third letter
U	UUU, UUC } Phe UUA, UUG } Leu	UCU, UCC, UCA, UCG } Ser	UAU, UAC } Tyr UAA, UAG } Ter	UGU, UGC } Cys UGA } Ter UGG } Trp	U C A G
C	CUU, CUC, CUA, CUG } Leu	CCU, CCC, CCA, CCG } Pro	CAU, CAC } His CAA, CAG } Gln	CGU, CGC, CGA, CGG } Arg	U C A G
A	AUU, AUC } Ile AUA AUG } Met	ACU, ACC, ACA, ACG } Thr	AAU, AAC } Asn AAA, AAG } Lys	AGU, AGC } Ser AGA, AGG } Arg	U C A G
G	GUU, GUC, GUA, GUG } Val	GCU, GCC, GCA, GCG } Ala	GAU, GAC } Asp GAA, GAG } Glu	GGU, GGC, GGA, GGG } Gly	U C A G

Fig. 2.17 The genetic code.

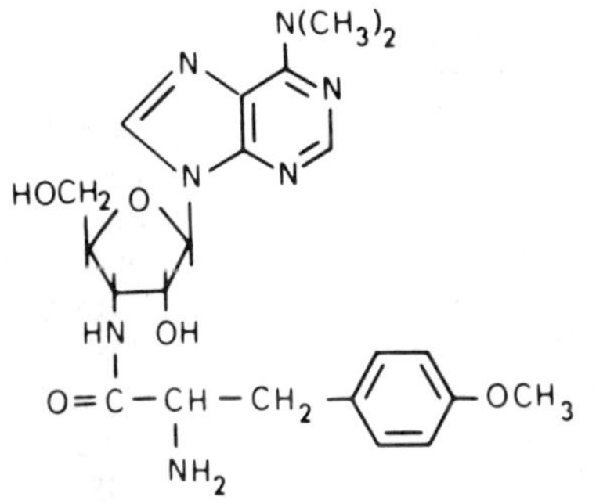

Fig. 2.18 The structure of puromycin.

released from the ribosome, because it lacks those parts of a tRNA molecule responsible for ribosome binding. The formation of amino acyl-puromycin and thus peptide bond formation can be detected by release of radioactive amino acid or peptide from the ribosome or by differential extraction of the puromycin derivative in organic solvents. Other antibiotics have also been of great value in studies of protein biosynthesis *in vitro*.

The puromycin-dependent assay of peptide bond formation can be further simplified by using as the substrate not amino acyl-tRNA but simply the terminal fragment containing the last few nucleotides at the 3′-end of the molecule before the ester bond with the amino acid. Thus, CAACCA-fMet which can be prepared by treating fMet-tRNA with RNAse T_1 (which cleaves to the 3′ side of G residues), can act as substrate in the so-called fragment reaction with puromycin [25]. The great advantage of this reaction is that since no tRNA binding is involved ribosomal subunits can be used. In this way the enzyme peptidyl-transferase was found to be located on the 50S ribosomal subunit [26].

After peptide bond formation the nascent peptide chain is attached to the tRNA in the A-site and the tRNA in the P-site carries no amino acid. The next step in protein synthesis is the relative movement of the ribosome along the messenger such that the uncharged tRNA is ejected, the peptidyl-tRNA is repositioned in the P-site and a new codon is exposed in the A-site (see Chapter 3 and Fig. 3.1). This relative movement (or translocation) is fundamental to the process of messenger read-out but it is extremely difficult to envisage the mechanics involved. The assay of translocation is not simple either but most methods measure changes in the state of the tRNA bound to the ribosome. Thus peptidyl-tRNA in the A-site or pre-translocation state is not released by puromycin because it is not present in the correct position relative to peptidyl transferase. After translocation peptidyl-tRNA is present in the P-site and therefore puromycin-sensitive. GTP is required to provide the energy required for translocation and GTPase activity is therefore sometimes used as an assay, but this is somewhat unsatisfactory as it is relatively non-specific. A more recent method to assay for translocation uses defined polymers of the kind AUG-UUU-UUU, and antibodies to purified elongation factors. At the end of the partial reaction the peptide made is identified (as either MetPhe or MetPhePhe) and the involvement of translocation assessed [27].

The initiation and termination reactions carried out by isolated ribosomes or ribosomal subunits are often modifications of the partial reactions mentioned above. For example initiation can be assayed by the binding of initiator tRNA to ribosomes, or subunits; termination by the release of fMet by hydrolysis of fMet-tRNA present in the P-site. The latter is catalysed by termination factors and requires termination triplets. As initiation of protein synthesis occurs on ribosomal subunits there must also be enzymes or factors to catalyse the dissociation of ribosomes prior to the formation of an initiation complex. These factors can be assayed by analysis of the resulting particles on sucrose gradients.

2.2.3 Structure and function

Recent progress in understanding structure-function relationships depends on the ability to isolate and separate large amounts of ribosomal RNA and each of the separate ribosomal proteins. The structure of the separate component is studied in detail. To determine their function a simple experimental approach is often taken; some component of the ribosome is changed and each of the partial reactions, described in detail above, is examined and an attempt made to correlate that change in structure with the change in properties. Such changes can be studied, for example by isolated ribosomes from well-characterized mutants and identifying the mutant ribosomal protein or RNA responsible for the altered functioning of the ribosome.

Readers will realize that research into the structure and function of the ribosome is one of the most active and rapidly moving fields in current molecular biology and attempts to describe the bacterial ribosome in terms of over 50 component parts. In a monograph of this size only a brief description of some of the approaches being used and a few specific results can be mentioned as examples; for further details readers are referred to more specialist treatments [28, 30, 31, 32, 34, 36, 37].

rRNA. Bacterial ribosomal RNA can be separated into 3 size classes, 23S (1.1×10^6 daltons), 16S (0.55×10^6 daltons) and 5S (40 000 daltons). Much early work on rRNA concentrated on details of its biosynthesis, the number of genes coding for rRNA in different organisms and detailed comparison of the molecular weights of rRNA from many different organisms. This data gives little direct information on the function of rRNA within ribosomes. More recent work has concentrated on the primary and secondary structure of rRNA and this should be of value in building up tertiary structures for rRNA within which the various proteins can be fitted.

Bacterial ribosomal RNA can be labelled with ^{32}P to very high specific activity and it is therefore suitable for nucleotide sequencing. The primary sequence of 5S RNA from several organisms has been determined, but as yet its tertiary structure is uncertain. 5S RNA is bound to the 50S ribosomal subunit and ribosomes reconstituted in its absence are defective in all functions. Three specific ribosomal proteins are required to bind 5S RNA to 23S RNA, two referred to as L18 and L25 bind to 5S RNA, the other, L6, binds to 23S RNA [28]. Almost all tRNA molecules contain a sequence GTΨCG and *E. coli* 5S RNA contains a partially complementary sequence, CGAAC. It is possible that 5S RNA is involved in binding to this common sequence and thereby attaching tRNA to the ribosome. This idea is supported by the finding that the tetranucleotide TΨCG binds to 50S subunits and by doing so inhibits elongation factor 1 dependent binding of amino acyl-tRNA and GTP hydrolysis [29].

5S RNA from many different sources has now been sequenced. It is estimated that *E. coli* has 6 genes for 5S RNA and some sequence differences have been detected at specific positions in the molecule. *Xenopus laevis* on the other hand has 24 000 copies of the 5S genes per haploid genome, and at least two different species have been isolated (one from oocytes, the other from somatic cells) and sequenced (Fig. 2.19). The two species are different in at least 5 positions in the primary sequence. It

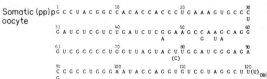

Fig. 2.19 Nucleotide sequence of *Xenopus laevis* 5S RNA (from [30]).

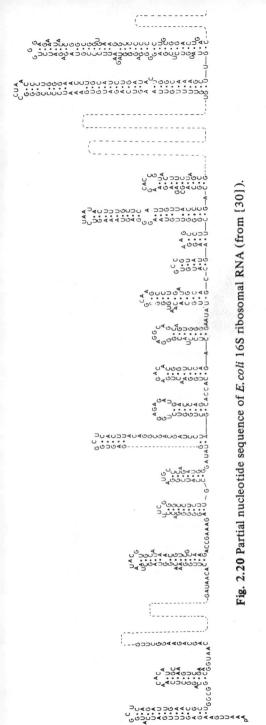

Fig. 2.20 Partial nucleotide sequence of *E. coli* 16S ribosomal RNA (from [30]).

remains to be seen whether the different 5S RNAs have different activities and as such represent a switch in the function of the ribosome during development. Similar developmental changes in other rRNAs or ribosomal protein components could also occur but so far the data to support this idea is not compelling.

The high molecular weight of 16S and 23S rRNAs from *E. coli* makes any kind of analysis difficult. Early work showed that the G + C content of rRNA is relatively high and that some modified nucleotides are present. The determination of the complete nucleotide sequence of rRNA is an exceedingly complex task, but as with mRNA specific portions can be examined more easily. For example, the 5′ and 3′ termini have been sequenced as have the regions around the methylated nucleotides. More extensive sequence data are now becoming available and this suggests hairpin type structures of varying lengths may exist within the rRNA (Fig. 2.20). Comparison with 23S RNA, for which there is much less sequence data, indicates very little sequence homology between 16 and 23S RNA but there may be considerable homology within the 23S rRNA itself.

Currently much attention is being given to the complete elucidation of the primary sequence of both species of bacterial rRNA. However, the sequences themselves are of limited value, particularly since it is unlikely that the secondary and tertiary structures will be deduced from the primary sequence. For this reason partial fragmentation of ribosomes with ribonuclease, followed by separation of the fragments and analysis of the proteins and RNA sequences present is of great value in attempting to integrate RNA sequence data with work on the isolated ribosomal proteins. Similar results can be obtained by first isolating fragments of rRNA and then attempting to rebind specific ribosomal proteins.

Ideas on the importance of the rRNA in the structure and function of ribosomes have undergone many fluctuations with time. Currently the view is that the bulk of the structure of the ribosome is RNA and water, with the proteins interdispersed in that basic structure. This model therefore predicts a more dynamic role for rRNA in protein biosynthesis than has previously been thought likely. One idea, for example, is that the 3'OH end of 16S rRNA plays a specific role in mRNA binding (see Section 3.3.3 p. 52). This model for ribosome structure is also consistent with the finding that ribosomal proteins are quite distant from one another and interact mainly with rRNA (see Section 2.2.3) and that the tRNA binding site on ribosomes is predominantly RNA (see Section 2.3.5).

Ribosomal proteins. Ribosomal proteins can be separated on an analytical scale by 2-dimensional polyacrylamide gel electrophoresis (Fig. 2.21). To isolate the proteins in quantities sufficient for protein sequencing, however, large amounts of ribosomal subunits must be isolated, usually by preparative zonal ultra-centrifugation. After disruption many of the proteins from the 30S subunit (which are all basic) can be separated by chromatography on cellulose phosphate and fractions containing more than one protein can be resolved into pure species by molecular sieving on Sephadex G100 or by rechromatography. The 50S proteins are too numerous for separation by a single chromatography step; they are therefore partially fractionated by differential salt extraction of the purified 50S subunit. This yeilds three fractions, each of which can be separated into pure protein fractions as described for the 30S proteins.

The proteins isolated from a given ribosome preparation are usually considered true ribosomal proteins if they remain attached to the particles after extensive washing with high salt solutions and if they are present in approximately molar yield. This definition must for the present remain merely operational because it is not yet certain whether some ribosomal proteins are only transiently present on the ribosome or if they vary during the growth cycle or with location within a cell. This latter view is supported by Kurland, who defines as ribosomal proteins those which are necessary for ribosomal function and whose removal impairs protein biosynthesis [32]. This definition would include all initiation, elongation and termination factors as ribosomal proteins.

There is general agreement that the *E.coli* small subunit contains 21 proteins (designated S1 to S21) and the large subunit 34 proteins (L1 to L34). Each of these is the subject of intense study and antibodies to each have been raised in rabbits [33]. All 55 ribosomal proteins are immunologically distinct, except for L7 and L12, which show cross reaction and have been shown by peptide analysis to be identical except for the presence of an N-acetyl group in L7; and S20 and L26 which cross react, and are present only in about half molar yield in their respective subunits. Possibly they are the same protein, which can exist on either

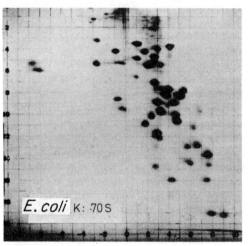

Fig. 2.21 Separation of ribosomal proteins by 2-dimensional polyacrylamide gel electrophoresis (from [31]).

subunit.

Antibodies have been of great value in identifying individual ribosomal proteins obtained in different laboratories, in relating antibiotic-resistant mutant ribosomal proteins, and in characterizing those proteins that will associate with isolated rRNA. The finding that all proteins present in the 30S subunit and most present in the 50S subunit are accessible to antibody also suggests that all are at least partially exposed on the surface of the ribosome.

The use of antibodies to ribosomal proteins has been refined by purifying the immunoglobin G from the antisera, and preparing monovalent Fab-fragments by digesting the 1gG with papain. This has the dual advantage of removing contaminating RNAse and protease acivity and creating a much smaller probe. This can then be used to characterize the particular proteins involved in any reaction (see for example their use in the characterization of L7 and L12 described below) without unduly interfering with other more distant ribosomal proteins.

Sulphydryl reagents such as parachloromercuri-benzoate, iodoacetate and N-ethylmaleimide have been used to modify ribosomes chemically, and the remaining activity and the exact proteins modified examined. Such methods are seldom completely specific and give relatively little information other than to suggest that certain proteins are present on the outside surface of the ribosome. More specific chemical reaction is possible however, for example modification of isolated ribosomal proteins S11 and S12 with tetra-nitro-methane prior to reconstitution gives particles lacking the ability to blind Phe-tRNA. This type of approach can be expected to be used more extensively in the future.

Bifunctional chemical reagents are valuable because not only do they give an indication of the function of particular proteins but also some idea of the spacial arrangements of ribosomal proteins [34]. Reagents, such as dimethyladipimidate, are chosen because they react with lysine residues in polypeptides and in addition are easily removed from the isolated pairs of cross-linked proteins, so that the identification of the modified proteins is possible. Such modification can be attempted either using whole ribosomes or subunits or isolated pairs of proteins prior to reconstitution. Results of this kind of analysis indicate that bifunctional reagents with an inter-group distance of 9Å makes relatively few ($\sim$ 6) cross links. Since each protein has approximately 15 lysine residues and monovalent reagents attack virtually every protein, this indicates that each ribosomal protein may not touch its neighbour. Instead each may be attached primarily to RNA and dispersed throughout the matrix of the ribosome. Model building suggests that if each protein is evenly spaced and roughly spherical adjacent proteins should be on average 8Å from a neighbour, and this is supported by the finding that if the inter-site distance in the reagent is increased to 12 Å the yield of cross linked proteins is greater, whereas if it is only 5Å no cross linking occurs. Data obtained using such reagents suggests, for example, that proteins S5 and S8 are adjacent on the small subunit.

The total reconstitution of ribosomal subunits from their separate component parts must rate as one of the most spectacular experiments of the last decade. Reconstitution of 30S particles, in addition to purified 16S RNA and all 21 proteins, displays an absolute requirement for Mg^{++} ions, and heat. Similar conditions are required for reconstitution of 50S subunits but in this case the need for native undernatured ribosomal proteins is of paramount importance. In fact the first reconstituion of 50S particles used components from *Bacillus stearothermophilus,* which seem to be unusually stable. The reconstitution technique is of immense value in attempting to identify the function of any particular protein, since ribosomes can be constructed *in vitro* either lacking the particular protein in question or containing a mutant form. Spatial relationships can be assessed by cross-linking two isolated proteins and then attempting reconstitution. Presumably only if the two proteins are physically close will reconstitution still be possible.

A further use of reconstitution in assessing spatial relationships comes from the study of the order of addition of each of the various ribosomal proteins, and this has led to the construction of an assembly map (Fig. 2.22). Proteins early in the map are thought of as

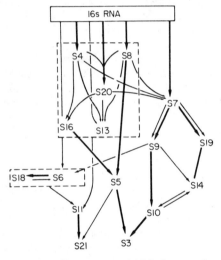

Fig. 2.22 Assembly map of 30S ribosomal proteins (from [35]).

necessary to fold the nascent particle into the correct conformation such that the later proteins may bind and they need not necessarily have any function in mature ribosomal particles. One such early protein is S9; particles reconstituted in its absence lack all functions of the 30S particle and sediment at 25S. On the other hand, partial reconstitution starting with 23S core particles, which are made by splitting off some proteins of the 30S subunit (including S9) by CsCl centrifugation, does not require the addition of S9. Presumably 23S cores retain the correct conformation and S9 is therefore not necessary.

Amongst the early proteins in the 30S subunit assembly are some that bind tightly to naked 16S RNA; these include S4, S7, S8 and S20. These bind to specific sites on the 16S RNA in a 1:1 molar ratio. Interestingly all the binding sites (except S7) are near to the 5'-end of the rRNA. The presence of the proteins protects large contiguous fragments of the RNA from nuclease attack, suggesting that the proteins bind to at least two sites which are distant from one another in the primary sequence of the RNA but are brought together either by the folding of the RNA or the presence of the proteins, or both.

The 30S subunit proteins can be further divided into four groups depending on the properties of particles reconstituted in their absence in displaying (1) a grossly altered sedimentation rate (S4, S7, S8, S9, S16 and S17) (2) a slight alteration in sedimentation rate but large decrease in functional activity (S3, S5, S10, S11, S14, S19); (3) no requirement in assembly but necessary or stimulatory for function (S2, S6, S12, S13, S18, S20 and S21) and (4) no effect on assembly or function (S1).

Reconstitution has also been achieved using RNA and protein from widely different organisms, indicating that some protein binding sites must be highly conserved, for example *B. stearothermophilus* 16S RNA and *E. coli* proteins give active hybrid subunits.

Ribosomal RNA can be fragmented by RNAse prior to binding ribosomal proteins or conversely, as mentioned above, subunits can be digested with nuclease and the protected fragments purified and analysed for specific RNA sequences and the presence of specific proteins. This approach again gives data about the spatial arrangements of protein and RNA, but so far ribosomal fragments isolated in this way have little biological activity and therefore give little data about function.

Many antibiotics act by interfering specifically with the activity of the ribosome. Puromycin has already been mentioned; other antibiotics inactivate other non-ribosomal factors required for protein synthesis and do not directly inhibit the ribosome itself, but many others have a direct influence on ribosomal proteins or rRNA. Streptomycin interacts with ribosomal protein S12 and thereby inhibits protein synthesis. Mutants which are resistant to streptomycin and others which are dependent on the antibiotic have been shown by reconstitution tests to have an altered S12 protein. Many other antibiotics have been shown to influence specific proteins (for example erythromycin, L22; spiramycin, L4). Unlike the antibiotics just mentioned, kasugomycin inhibits ribosome function not by interacting with ribosomal proteins but by inhibiting methylation of rRNA. These few examples serve to illustrate the value of antibiotics in studying ribosomal structure and function; for much greater detail see one of the specialist texts [36].

In the previous discussion the ribosomal components have been described in relation to one another and somewhat in isolation from other non-ribosomal factors required for protein synthesis. However similar structure/function relationships can also be deduced for these components, for example the site on the ribosome to which tRNA molecules attach can be established by chemically cross-linking modified tRNA species and ribosomal proteins, followed by the identification of the cross-linked molecules. Analyses of this kind are mentioned later (see Sections 2.3.5 and 2.5).

Proteins L7 and L12. In the previous section details of some of the methods used to probe the structure and function relationships of ribosomal proteins have been described in general terms. In this section some detailed results obtained with the closely related proteins L7 and L12 will be presented as an example of the progress that has been made in assigning specific functions to a particular ribosomal protein and in defining its structure. L7 and L12 are not necessarily typical ribosomal proteins but they are of great importance, and have been the subject of considerable study.

L7 and L12 each have a molecular weight of about 13 000 daltons and are both acidic proteins, with a high content of alanine (24%) and glutamic acid and glutamine (15%). They contain no histidine, cysteine or tryptophan and only one arginine and three lysine residues. The α-helical content of the proteins is very high and estimated to be 50–60%. Antibodies against one of the two proteins completely cross-react with the other and *vice versa*; the only difference so far detected between the two proteins is that L7 starts N-acetyl serine whereas L12 starts with free N-terminal serine. The proteins occur in at least two and often three copies per ribosome. The sum of the amount of both proteins present in *E. coli* varies with growth media and the ratio between the two proteins varies with growth conditions. The chemical and physical properties of L7 and L12 resemble closely those of contractile proteins such as myosin and flagellin. L7 and L12 from widely differing bacterial species show high cross reactivity with antibody to the *E. coli* proteins and can form active hybrid ribosomes with *E. coli* cores.

50S ribosomal particles lacking L7 and L12 lack GTPase activity but elongation factor G dependent GTPase can be restored by addition of either protein. Similarly antibodies to L7 and L12 added to a reconstitution mixture inhibit the GTPase activity of the particles formed under these conditions. Furthermore, the formation of a complex (EFG:GDP:50S:FA) between ribosomes, elongation factor G and GDP which is stabilised by an inhibitor of translocation called fusidic acid (FA) is prevented by Fab fragments of antibodies to L7 and L12. All these properties suggest that L7 and L12 are the ribosomal proteins involved in GTP hydrolysis at translocation; perhaps these molecules provide the 'muscle' to move the ribosome along the mRNA. Further experiments are continuing to learn more of the molecular mechanisms involved. For example specific modification of the proteins at known positions (e.g. by modification of the single arginine residue, or the three lysine residues, or by CNBr cleavage at the methionine residues), followed by functional tests should reveal which portions of the molecule are important in say GTPase activity and which in binding to the ribosome. So far attempts to bind free GTP to the isolated proteins have failed, but there is evidence to suggest that other ribosomal GTPase reactions catalysed by elongation factor T and initiation and termination factors also occur at sites close to the elongation factor G GTPase site. This perhaps indicates L7 and L12 have several roles in protein synthesis. So far it is not known which of the closely related proteins is the more active *in vivo* nor is the change in the relative amounts of each understood. Perhaps L7 and L12 form some site which, in addition to functioning in translocation, is also involved in initiation, elongation and termination factor dependent reactions. This is discussed again below. In an attempt to define further the L7/L12 ribosomal site 50S subunits lacking these two proteins have been reconstituted and reacted in turn with antibodies to each of the proteins present in the particle. Antibodies to L6 and L18 interfere with the subsequent addition of L7 and L12 and presumably therefore these proteins are also among those close to the 'translocation' site.

2.2.4 *Biosynthesis*

The biosynthesis of ribosomes is described in detail in the monograph by Dr. Burdon. However, the biosynthesis of ribosomes, at least in bacteria, is important in assessing translation control mechanisms because cellular growth rate and macromolecular synthesis are intimately related to the number of ribosomes per cell [37]. In *E.coli* there are about 6 genes coding for the rRNA species but as mentioned earlier there is no good evidence for variation in the genetic content of different genes. All three rRNA species are clustered, in the order 16, 23, 5 and there is only one promotor for their synthesis. In mutants lacking RNAse III there is evidence for a giant precursor rRNA molecule containing all three species, and each of the stable species has intermediate precursor forms that have been characterized. The 16S precursor, for example, has additional sequences at both ends. The ribosomal protein genes of *E.coli* are also partly clustered and there is evidence that the proteins are synthesized co-ordinately (possibly as a polycistronic mRNA) and synthesis is under stringent control.

Little is known about the details of assembly of nascent ribosomes, but it is evidently a complex reaction. Many ribosome assembly mutations are known, and these were of importance in the early work on ribosome reconstitution. *In vitro* reconstitution can be attempted using rRNA precursors, but as might have been predicted from the reconstitution assembly map a definite sequence of addition is necessary, and so far such particles are inactive. It is known for example that premature addition of protein S4 inhibits cleavage of pre-16S to mature 16S. Methylation of some bases of rRNA, which occurs after addition of the early binding proteins, is also a problem.

Ribosome biosynthesis in higher cells has also been studied in detail. Synthesis and maturation of the high molecular weight rRNA occurs in the nucleus and involves a well established pattern of cleavages. However, the exact number of genes and chromosomal location varies with different organisms.

2.3 Transfer RNA

2.3.1 Discovery and properties

The existence of tRNA was predicted by Crick in the mid 1950's. The sequence hypothesis mentioned earlier dictated that the linear sequence of nucleotides in DNA is translated into the linear sequence of amino acids in a protein but at that time nothing was known of the components involved, and attempts to explain the specificity of such a relationship in terms of ionic or hydrophobic interactions between nucleic acids and amino acids seemed implausible. A hypothetical 'adaptor molecule' was invoked which was composed of nucleic acid and on the one hand could activate a specific amino acid and on the other hand could recognize signals in nucleotide sequences and interact with them through H-bonding. Subsequent investigations into the adaptor molecules which later became known as soluble RNA (sRNA) and later still as transfer RNA (tRNA) have firmly established many of Crick's early predictions, and also his later modification 'the wobble hypothesis'.

tRNA has now been isolated from a wide variety of different organisms; it is easily purified from ribosomal RNA because it is soluble in high concentrations of salt. tRNA is characterized by a high content of modified nucleotides (minor bases) and it has a molecular weight of about 25–30 000 daltons corresponding to a molecule of between 75–90 nucleotides. Enzymes which activate amino acids have also been isolated from a variety of sources, they catalyse the reaction:

amino acid + ATP $\rightleftharpoons$ amino acid: AMP + PP
amino acid: AMP + tRNA $\rightleftharpoons$ amino acyl-tRNA + AMP

whereby a high energy ester bond is formed between the carboxyl group of the amino acid and the free 3′ hydroxyl group of tRNA. The structure and function of tRNA have been the subject of an immense amount of work by molecular biologists. This is not only because of the important role that amino acyl-tRNA plays in protein biosynthesis, but also because

Anticodon	Codon
A	U
C	G
G	U C
I	U C A
U	G A

Fig. 2.23 The wobble hypothesis (from [38]).

these molecules interact with a wide variety of other components (such as synthetases, ribosomes, mRNA, elongation factors etc.) and are therefore an excellent model system for the study of protein: nucleic acid interactions.

It was at first thought that there were twenty different tRNA species, one for each amino acid. With the discovery of the degenerate nature of the genetic code it was suggested that there might be sixty-four tRNA species, one for each possible codon. As isolation and purification techniques improved it was found that in reality there are an intermediate number of tRNA molecules, perhaps 50 in *E. coli* and that the anticodon region of some tRNA species is able to recognize more than one codon in mRNA. This observation was given a firm theoretical basis in the wobble hypothesis which predicts that by a slight distortion (or wobble) in the usual pattern of Watson-Crick base pairing one base in an anticodon can interact with as many as three different bases in the 3rd position of the codon (Fig. 2.23). Thus inosine (I) in the anticodon can base-pair with U, C or A and in this way tRNA^{Ala} with the anticodon IGC can read three alanine codons, GCU, GCC and GCA. There are probably only twenty amino acyl-tRNA synthetases and these enzymes must therefore have the ability to recognize more than one tRNA species.

Experimental proof that tRNAs are indeed the adaptor molecules they were predicted to be came in an elegant experiment in which the -SH group of the cysteine residue of Cys-tRNA^{Cys} was reduced by H_2/Ni to -H thereby generating Ala-tRNA^{Cys}. The tRNA molecule with the modified (i.e. incorrect) amino acid attached donated the alanine in response to cysteine codons present in synthetic mRNA added to a cell-free system [39]. Thus the adaptor molecule is responsible for reading the codons in mRNA and the amino acid attached to the adaptor is irrelevant. Recently similar experiments have been possible using mischarged amino acyl-tRNA, obtained by charging tRNA from one species with synthetase from another organism. Sometimes such heterologous charging leads to specific mischarging; but again the tRNA decides the specificity of codon reconition, not the incorrect amino acid.

2.3.2 Reactions of tRNA

tRNA plays a central role in protein biosynthesis and interacts with a very wide range of other molecules. Many of the reactions are referred to in detail in other sections, but they are briefly

---

RNAse cleavage of precursor
Methylation/modification
CCA addition
Amino acylation
Formylation of initiator tRNA
Elongation factor I
Anticodon/codon
Ribosome binding
Elongation factor II

Fig. 2.24 Some reactions involving tRNA.

summarized here as a reminder of the many enzyme recognition sites that must be present in tRNA (Fig. 2.24). In the later sections an attempt is made to correlate these properties with the known structure of tRNA molecules.

tRNA is made in precursor form in both prokaryotes and eukaryotes; specific RNAses which cleave the precursors and other enzymes that specifically modify certain nucleotides in nascent tRNA molecules have been identified and their activity studies *in vitro*. As isolated from some organisms, tRNA lacks the terminal sequence CCA, and another enzyme, nucleotidyl transferase, has been characterized which specifically replaces this sequence.

The amino acylation (or charging) reaction can also be studied in highly purified cell-free systems. Radioactively labelled amino acid is used and charging detected by precipitating the amino acyl-tRNA with cold TCA. Charged tRNA can be isolated by extraction with phenol and used in cell-free protein synthesis, or in ribosome binding studies using either synthetic or natural mRNA. The detailed enzymology of the binding of $a\alpha$-tRNA to ribosomes has been investigated using purified elongation factors.

The multiplicity of these reactions involving tRNA, some of which are specific to an individual species (e.g. methylation of minor bases), some specific to a small class of molecule (e.g. charging) and others which are common to almost all (e.g. ribosome binding) implies that such molecules must have some basic overall shape that is common to all species upon which the specific characteristics of individual species are superimposed. The structure of tRNA would be expected to reflect these characteristics.

2.3.3 Purification and primary sequence of tRNA

Large quantities of tRNA can be prepared from micro-organisms, but initially a major difficulty

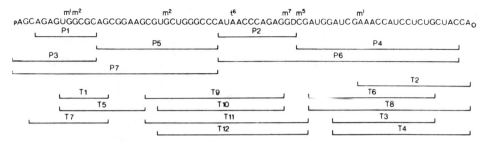

Fig. 2.25 Derivation of primary sequence of tRNA^{Met} by overlapping partial fragments (from [40]).

in determining the structure of a single species of tRNA lay in purifying one particular molecule from a mixture of fifty or more very closely related molecules. Early techniques of sequencing nucleic acids needed optical density amounts of material and therefore large scale fractionation methods such as counter current distribution had to be used. The development of radioactive methods by Sanger meant that much smaller quantities were necessary and ion exchange and adsorption chromatography on columns have been utilized. Nowadays resins of remarkable resolving power (such as DEAE-Sephadex, bensoylated DEAE-cellulose and RPC-5) make the purification of specific species of tRNA a routine laboratory procedure. Nevertheless, the need for very large quantities of highly purified tRNAs has arisen again for use in experiments involving the crystallisation of tRNA for physical studies; increasingly large scale column fractionation is being used for this purpose.

Classical sequencing methods involve cleaving a tRNA molecule into small oligonucleotides using specific ribonucleases (such as RNAse T_1) and separation of the oligonucleotides by column chromatography. The primary oligonucleotides are next cleaved with a second enzyme (such as pancreatic RNAse to cleave after U and C) and the small oligonucleotides separated on paper. In this way the order of the bases in the primary oligonucleotides can be deduced. Overlapping the primaries is achieved using a second specific enzyme to cleave the whole tRNA molecule and using partial cleavage conditions (Fig. 2.25). Sequencing is made simpler because many minor bases are present in tRNA and these act as internal markers. Nevertheless this method is somewhat laborious and time-consuming. Radioactive techniques are much quicker because many of the oligonucleotide separation procedures are performed by two-dimensional electrophoresis on paper and detection utilises autoradiography (Fig. 2.26).

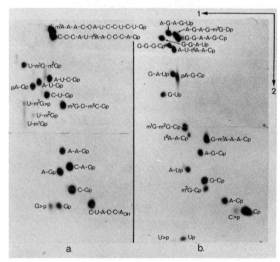

Fig. 2.26 Fingerprints of mouse tRNA^{Met} (a) T_1 RNAse (b) pancreatic RNAse (from [40]).

31

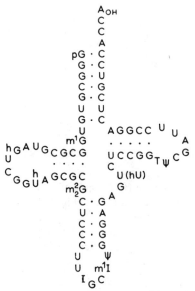

Fig. 2.27 Sequence of yeast tRNA^Ala (from [40]).

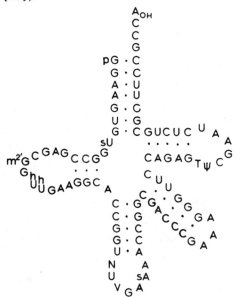

Fig. 2.28 Sequence of *E. coli* tRNA^Ser (from [40]).

Using essentially the same strategy of enzymatic degradation, sequencing tRNA molecules on the radioactive scale has now become very simple, and over fifty such molecules have been sequenced in this way. A modification of the radioactive method that can be applied to systems that are difficult to label *in vivo* is the enzymatic introduction of ^{32}P into oligonucleotides by specific kinases and ^{32}P-ATP after ribonuclease cleavage of unlabelled tRNA.

All the tRNA molecules sequenced to date appear to fold into the so-called clover-leaf model with a stem, three large arms consisting of a stem and a loop, and sometimes an extra arm (Fig. 2.27 and 2.28). At various times the loops and stems have been given different names; one currently-accepted nomenclature is shown in Fig. 2.29. Certain features of the primary sequence appear to be constant. Thus the 3′ terminal sequence -CCA$_{OH}$ and the sequence TΨC in loop IV are ubiquitous. The nucleotide to the 5′ side of the anticodon is always U and that to the 3′ side is a modified purine. Perhaps these two characteristics are related to the recognition of the anti-codon by mRNA and the need to accommodate 'wobble'. This normally occurs at the 3′-end of the codon (i.e. adjacent to the U); but in the case of the initiator tRNA (see Section 2.4) wobble occurs at the 5′-end and in this case the base adjacent to the anti-codon at the 3′-end is not modified. Other constant nucleotides can be seen in Fig. 2.29. The nucleotides present in the stem regions are extremely variable but the number of base pairs per stem is fairly constant. Stem b can have either 3 or 4 base pairs and stem d can have anything from zero to seven base pairs. These differences fall into three groups and form the basis of a simple classification of tRNAs (Fig. 2.30).

It was at first hoped that examination and comparison of the primary sequences of several tRNAs would indicate which portions of the molecules were involved with each

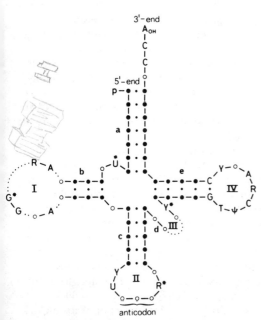

Fig. 2.29 Generalized clover leaf structure showing constant features (from [40]).

particular function. It was expected that some features would be common (e.g. the ribosome binding site) whereas others would determine specificity for say amino acyl RNA synthetase recognition. Some progress along these lines has been made, for example TΨCG probably represents the common ribosome binding site of

Class I	4 base pairs in stem b
subclass A	5 bases in loop III including m^7G
subclass B	5 bases in loop III no m^7G
subclass C	3 or 4 bases in loop III
Class II	3 base pairs in stem b
	3–5 bases in loop III
Class III	3 base pairs in stem b
	13–21 bases in loop III

Fig. 2.30 Classification of tRNA.

tRNA (see Section 2.2.3 p. 22). However, comparison of the primary sequence of several tRNAs that can be charged by the same enzyme does not immediately suggest a synthetase recognition site and shows that the relationships between primary structure and enzymatic function are not obvious. For example the two Met-tRNAs of *E. coli*, which are charged by the same enzyme (see Section 2.4) are as different from one another as any two randomly chosen tRNAs. Probably the reason that we are unable to pick out enzyme recognition sites is that the primary sequences reflect structures required in maintaining tertiary structure, as well as others required in determining specific interactions with other molecules. An understanding of the parts of the molecule involved in maintaining the tertiary structure should therefore allow a distinction to be made between sequences required for structure and those defining specificity. In addition it should suggest an overall shape for the molecule which might give some clue as to how it functions.

2.3.4 Tertiary structure of tRNA

Some idea of the tertiary structure of tRNA is gained during the determination of the primary sequence. Ribonuclease digestions, carried out under conditions which allow cleavage of only a few phosphodiester bonds and thereby give overlap data, attack first those bonds which are most exposed to the enzyme. If the digestion conditions are such that the tertiary structure of the tRNA is conserved, some structural data can be deduced.

Similar information can be obtained by chemically modifying tRNA under relatively mild conditions and examining which nucleotides have been attacked. Such nucleotides are presumed to be exposed in the molecule and not involved in H-bonding. Reagents such as methoxyamine, which specially reacts with C residues; carbodiimide derivatives which react with U and G; perphthalic acid which reacts

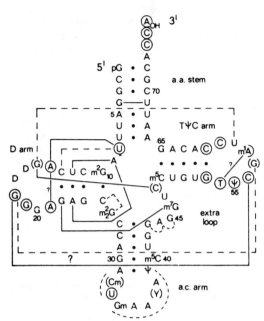

Fig. 2.31 Structure of tRNA^{Phe} showing additional interactions involved in maintaining tertiary structure (from [42]).

The definitive method of determining the tertiary structure of tRNA is X-ray crystallography. The major problems in applying this technique have been firstly making suitable stable crystals and secondly obtaining good isomorphous derivatives. Indeed it has taken over six years from the first crystallization of tRNA to the partial solution of the tertiary structure of tRNA^{Phe}.

Fig. 2.31 shows the clover leaf model of yeast tRNA^{Phe} showing the additional interactions present in the tertiary structure, as determined by X-ray crystallography and reported in 1974 [42]. Fig. 2.32 shows a schematic diagram of the three-dimensional structure; the molecule is roughly T-shaped with the a and e stems forming a long helical region (the cross number of the T) and the b and c stems forming a second helical region (the upright of the T). The shape of the molecule is maintained by base pairing, some involving non-Watson-Crick interactions (e.g. positions 15 and 48 form a reverse Watson-Crick base pair) and others involving 3 bases (e.g. positions 8, 14 and 21). Additional stability is given by stacking forces (e.g. in the anticodon loop). A comparison of the sequence of yeast tRNA^{Phe} with that of other tRNA species suggests that many tRNAs (especially those in class I (Fig. 2.30)) could be accommodated within the overall three-dimensional shape established for tRNA^{Phe}. This suggests that many tRNA molecules will have a similar shape and allows us to generalize about those portions of the tRNA molecule involved in maintaining the structure.

2.3.5 Structure and function of tRNA

As discussed above, primary sequence analysis shows that some portions of tRNA molecules are invariant (Fig. 2.29). Comparison of these regions with the crystal structure of tRNA^{Phe} shows that several of these invariant regions may be crucial in holding together the tertiary structure. For example in all tRNAs so far sequenced

with A and many others have been used. Often modification conditions are harsh and non-physiological, and introduction of the modified groups probably disrupts the tertiary structure; nevertheless such techniques have been useful and show that regions in the single-stranded portions of the clover-leaf model interact faster than the helical regions. In addition, nucleotides which are present in the single-stranded regions also react to different degrees and at different rates and this is presumably due to partial protection caused by the tertiary structure. For example the C residues of yeast tRNA^{Phe} contained in the sequences -CCA react more readily with methoxyamine than do any of the other C residues. Such patterns of chemical reactivity must be taken into account when assessing the validity of three-dimensional models for tRNA deduced by other methods [41].

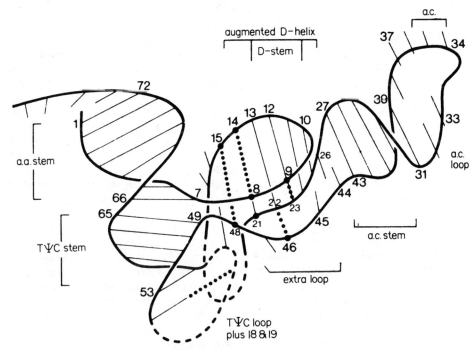

Fig. 2.32 Tertiary structure of tRNA^{Phe} (from [42]).

(except one) there is a purine residue at position 15 and a pyrimidine at postion 48. These two residues are involved in a reverse Watson-Crick base pair which cannot be readily accommodated in the model if the purine is in position 48 and the pyrimidine in position 15. This therefore strongly suggests that these two bases are involved in maintaining the tertiary structure in all tRNA species and consequently that all tRNAs have basically the same shape. This would satisfy the requirement that certain structures of all tRNAs are in the same relative position i.e. the distance from anticodon to amino acid, and to the ribosome binding site (TΨCG) should be constant.

In addition to the common sites and sequences of the tRNA model just described, variant regions which are exposed in the tertiary structure are also present. Such regions may form the specific enzyme recognition sites. For example positions 16, 17 and 20 of loop 1 are both variable and exposed, and there is some evidence to invoke this region as a synthetase recognition site (see Section 2.5). The extra loop (loop III) is also extremely variable and exposed and this is probably another such discriminatory site. Undoubtedly, as the detailed tertiary structure of other tRNA molecules is established and experiments to establish the function of tRNA are related to the three-dimensional models, a clearer picture of enzyme recognition sites will rapidly emerge.

tRNA also plays a dynamic role during protein synthesis and attempts to correlate this function with the three-dimensional model are also in progress. It seems clear that at certain stages during either ribosome binding and/or translocation, conformational changes in the shape of tRNA occur. For example, as mentioned, the TΨCG loop is thought to interact directly with 5S RNA, yet this sequence is buried within the tertiary structure of tRNA in solution. Furthermore by changing the Mg^{++} concentration some tRNAs can be reversibly interconverted between active and inactive forms. It is therefore important that the structure of tRNA can accommodate some relative movement of its component parts.

In this respect it is of particular interest that the tRNA model contains a possible 'hinge' between the anticodon stem and the loop I helix. During translocation mRNA (with peptidyl-tRNA hydrogen bonded to it) is moved relative to the ribosome such that a new codon is exposed in the A-site of the ribosome. Perhaps the conformation of the anticodon region of the peptidyl-tRNA, which is normally stacked, changes at this time. Possibly a flip-flop type movement occurs around the hinge. In this way a rachet-type mechanism for translocation can be envisaged.

Whilst much effort in tRNA research is now concentrated on X-ray crystallographic methods, other avenues of investigation are being explored. Nuclear magnetic resonance studies detect secondary structure hydrogen bonded proton resonances. These can be assigned to different base pairs in the tertiary structure, thus providing a powerful new tool for structural analysis.

Chemical studies are also continuing, in attempts to correlate the new structural data available for tRNA with similar data for other cellular components with which tRNA interacts. For example, tRNA molecules with chemically or photo-reactive groups attached at known positions have been used to form cross-linked complexes with synthetases (see Section 2.5), elongation factors and ribosomes. It is most interesting that after binding to the P-site of ribosomes some probes (for example S (p-azidophenacyl) valyl-tRNA with the photo-affinity group attached to the 4-thio-U residue) link exclusively to the 16S RNA. Other similar reagents react with 23S RNA. This emphasizes the argument made earlier, that much of the ribosome is composed of RNA with proteins rather widely dispersed from one another. It is consistent with the observation that tRNA also interacts with 5S RNA. Perhaps the tRNA-binding site on ribosomes is composed mainly of RNA.

Other tRNA derivatives have been used as probes for ribosomal tRNA binding sites. p-nitrophenoxycarbonyl-^{35}S methionyl tRNA$_F$ can be used in initiation complex formation. The active ester interacts with ribosomal proteins and these can be identified by immunoprecipitation with specific antibodies to ribosomal proteins (see Section 2.2.3 p. 25). In this way proteins L27 and L15 were identified as possible constituents of the ribosomal binding site for initiator tRNA. Using different components of the binding reaction (e.g. replacing GTP or using antibiotics) the proteins involved in or near to other binding sites could also be identified, for instance peptidyl transferase.

From the previous sections it should be clear that our understanding of the molecular involvements in the different activities of tRNA has reached a fairly advanced stage. The aim of all research into the molecular biology of protein biosynthesis is to establish an equally clear understanding of all the reactions involved, but it will be some time before the ribosome or the translocation reaction is understood in such detail.

2.3.6 Mutant tRNAs

Previously (Section 2.3.1) experiments were described in which the amino acid attached to a specific tRNA molecule was changed or modified without affecting the ability of the tRNA to donate the modified amino acid in response to its original anticodon. In the same way a mutation in a tRNA gene can cause an alteration in the anticodon without necessarily altering the ability of the mutated tRNA to be charged with the correct amino acid. When this happens, although it can lead to the insertion of incorrect amino acids into polypeptides, this is not necessarily lethal as bacteria often have more than one gene coding for each tRNA.

It sometimes happens that a bacterium contains a second mutation in a readily identified protein that is caused by a single base change in the mRNA leading to the insertion of the wrong amino acid and consequently an inactive protein. Sometimes such mutations can be suppressed by a mutant tRNA, such as that described above, which reads the mutated codon as if it coded for the original correct amino acid. Many such missense suppressors have been identified.

A further class of suppressors are called nonsense suppressors. They are mutants tRNAs which are able to read termination codons and insert a particular amino acid. Two classes of such suppressors had been identified earlier by genetic experiments and termed amber and ochre suppressors. These are now known to be tRNAs which recognise codons related by one base change to the termination codons. Thus the SU_{III}^+ suppressor was identified as a tyrosine tRNA in which the anticodon had mutated from its normal G*UA (which reads the tyrosine codons UAU and UAC) to CUA which therefore is able to read the termination codon UAG. Other mutant tRNAs have been isolated; one class is interesting, because it is mutated in regions of RNA found in the precursor but not in mature tRNA. Such mutants are defective in cleavage and tRNA maturation. Other tRNA mutants have altered patterns of amino acid charging. A further very interesting suppressor tRNA is a mutated $tRNA^{Tryp}$ which recognizes UGA termination codons, not because it has a mutation in the anticodon, but because a base is changed in the stem of loop 1. All of these mutants are helpful in correlating the structure and function of tRNA molecules [43].

2.4 Initiator tRNA

Because synthetic mRNAs can be translated in cell-free systems it first seemed possible that ribosomes attached to the 5'-end of mRNA and moved along it until they fell off the other end. We now know that the conditions used for translation of synthetic polymers allow incorrect initiation to occur, and that natural mRNAs contain a specific initiation codon, which is recognized by the initiator tRNA.

The first indication of the existence of an initiator came with the discovery in *E.coli* of formyl methionyl-tRNA. Since the blocked amino group prevents fMet-tRNA donating amino acid into internal positions of polypeptides, a possible role in initiation was suggested. Soon after the nascent coat protein of phage f2 was shown to contain N-terminal formyl methionine, whereas the mature protein contains alanine. It is now clear that initiation in all organisms involves a related methionine tRNA species, and that all proteins contain an N-terminal (formyl) methionine, albeit in many cases only transiently.

At least two species of tRNA that can be charged with methionine are present in all organisms; one is the initiator tRNA, and in prokaryotes and mitochondria and chloroplasts it contains a formyl group on the amino group of the methionine residue; it is designated $tRNA_F^{Met}$. The other species donates internal methionine residues and is designated Met-$tRNA_M^{Met}$. Both bacterial tRNAs are amino-acylated by the same enzyme, but only $tRNA_F^{Met}$ interacts with the transformylase enzyme. Both tRNAs read AUG as a codon, but in addition $tRNA_F^{Met}$ recognizes GUG and UUG. Such a 5' wobble in codon recognition is very unusual. The binding of Met-$tRNA_F$ to ribo-

	$Met\text{-}tRNA_F$	$Met\text{-}tRNA_M$
Methionine tRNA synthetase	+	+
transformylase	+	−
codon AUG	+	+
GUG	+	−
UUG	+	−
initiation factor F_2	+	−
elongation factor T	−	+
puromycin reaction	+	−
binds to	30S	70S
donates methionine	N-terminal positions	internal positions

Fig. 2.33 Comparison between $tRNA_F^{Met}$ and $tRNA_M^{Met}$.

somes is catalysed by an initiation factor whereas binding of Met-tRNA$_M$ is catalysed by elongation factors. fMet-tRNA is unable to donate methionine into internal positions because it cannot form a stable complex with elongation factors even if the formyl group is not present. Conversly Met-tRNA$_M$ is not recognized by initiation factors even if it is formylated. The two tRNAs bind to the ribosome at different sites as judged by the reactivity of the bound tRNAs with puromycin. Thus a pattern of properties of the two tRNAs emerges (Fig. 2.33). Some properties are shared, whereas others are unique to the initiator tRNA, and suggest that this species has a structure uniquely adapted for its role in initiation [44].

When the primary sequences of the two E. coli $tRNA^{Met}$s are compared with one another (Fig. 2.34, 2.35) and with other tRNAs, a few structural features seem unique to the initiator whereas others appear common to all. Thus the 5' terminal nucleotide pGp of the initiator tRNA is not base paired, $tRNA_F^{Met}$ does not contain the common sequence GTΨCG but instead has GTΨCA and the base adjacent to the 3' side of the anticodon GAU is an unmodified A, in contrast to the modified base normally found in this position in other tRNAs. Thus, in all these respects the prokaryotic initiator appears structurally unique.

Ribosomes can be classified into 70S and 80S types. These are defined not only by approximate sedimentation value but also sensitivity to antibiotics such as chloramphenicol (70S) and cycloheximide (80S). All systems which contain 70S type ribosomes also contain fMet-tRNA$_F$. Met-tRNAs from cells containing 80S type ribosomes are not formylated in vivo, nevertheless a similar methionine accepting tRNA functions as the initiator tRNA in these systems and shares many of the properties of its prokaryotic counterpart. Thus a species of Met-tRNA (designated $tRNA_F^{Met}$) from yeast and mouse can be formylated in vitro by the E. coli enzyme transformylase. $tRNA_F^{Met}$ from both species is recognised by E. coli Met-tRNA synthetase, whereas $tRNA_M^{Met}$ is not [45]. However, when the primary sequence of all the different $tRNA^{Met}$ species are compared (Fig. 2.34-2.37) it is not at all obvious which structural feature acts for example as the common transformylase recognition site, because the unique features of the bacterial initiator are not retained in the eukaryotes. One unusual feature of the initiator from mouse is the sequence in loop IV which is normally GTΨCG and which is supposed to be involved in binding tRNA to ribosomes. Since the initiator binds to a different site on ribosomes perhaps it is not surprising that this region of the molecule is not conserved.

2.5 Amino acyl-tRNA synthetases

Amino-acyl-tRNA synthetases have been isolated and purified from a wide variety of organisms. The molecular weight and subunit structure appear to vary widely; for instance

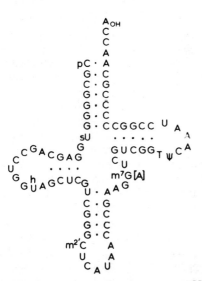

Fig. 2.34 Primary sequence of *E. coli* tRNA$_F^{Met}$ (from [44]).

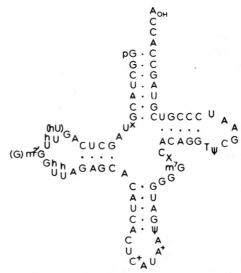

Fig. 2.35 Primary sequence of *E. coli* tRNA$_M^{Met}$ (from [44]).

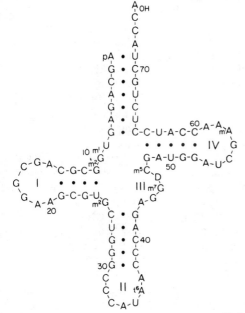

Fig. 2.36 Primary sequence of mouse tRNA$_F^{Met}$ (from [40]).

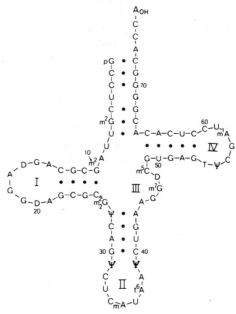

Fig. 2.37 Primary sequence of mouse tRNA$_M^{Met}$ (from [46]).

39

the *E. coli* isoleucine enzyme has a single polypeptide chain of molecular weight 118 000 whereas the methionine enzyme has a molecular weight of 180 000 and is composed of two identical subunits.

It is possible however to devise a relatively simple classification of synthetases if it is assumed that synthetase molecules are composed of different multiples of a 50 000 structure (see Fig. 2.38). Thus the simplest synthetase might be considered to be the yeast tyrosine-enzyme with molecular weight about 50 000 and structure α, the yeast tryptophan enzyme on the other hand is about 100 000 daltons with two subunits i.e. α_2. There is good evidence that some synthetases contain subunits with a single polypeptide which contains a large proportion of repeated sequence. Thus the *E. coli* methionine-synthetase might be considered to have the structure α'_2 (roughly 200 000 daltons) where α' itself is a gene duplicate of α.

It is probable that any given organism contains only one molecular species of amino acyl-tRNA synthetase for each amino acid. This has been shown by three different types of experiment; in several cases pure enzyme preparations have been shown to charge all the species of tRNA for a given amino acid; bacteria with mutant synthetase do not contain corresponding isoenzymes and antibodies to purified enzymes totally inhibit charging activity for the appropriate amino acid in crude bacterial extracts.

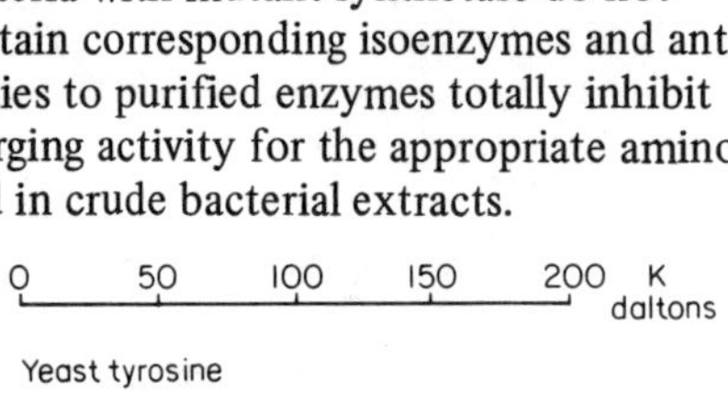

Fig. 2.38 Classification of amino acyl-tRNA synthetases by molecular weight.

There has been considerable interest in understanding the recognition patterns between the synthetase enzymes and their substrates. The site of interaction on the tRNA molecule (see Section 2.3) has been mentioned above and to date has mainly been deduced by comparing the primary sequences of tRNA molecules charged by the same enzyme, and of mutant, chemically altered and mischarged tRNA species. Such analysis is only suggestive and so far has not unequivocally identified the synthetase binding site. A more direct approach involves the formation of complexes between tRNA and different synthetases, followed by photo chemical cross linking by the action of ultraviolet light. The cross-linked regions of both the tRNA molecule and synthetase can be identified by the appropriate fingerprinting techniques. Such analysis suggests, for example, that *E. coli* ileu synthetase directly interacts with the loop I region of $tRNA^{Ileu}$.

A further elegant experiment to determine the tRNA binding site on the synthetase molecule used a reactive group attached to the amino acid moiety of Met-tRNA to attack the active centre of *E. coli* Met-tRNA synthetase. Thus a p-nitrophenoxycarbonyl-moiety was shown to react specifically with a lysine residue of the synthetase. A unique amino acid sequence around the tRNA binding site could therefore be determined [47].

Synthetases are of interest to enzymologists for a variety of reasons. First, they undergo a two-stage reaction involving the formation of an enzyme: amino acyl adenylate intermediate followed by binding of the tRNA and the charging of the amino acid onto the tRNA. Secondly many of the enzymes have two active centres and therefore a large number of binding sites (i.e. for tRNA, amino acid, ATP and Mg^{++}), each of which can be studied. Finally, the enzymes display a remarkable degree of specificity of reaction, in spite of the fact that their substrates, the amino acids and tRNAs, are all closely related. One interesting result from studies on the charging reaction is that in some cases the amino acid is first added to the 2'OH of the terminal adenosine residue and subsequently transferred to the usual 3' position. Further details of the interaction between tRNA and synthetases will become clearer when the complete amino acid sequence of a synthetase is determined, and when detailed X-ray crystallographic data on synthetases, and tRNA-synthetase complexes becomes available. It should be emphasized at this point however that the interaction of tRNA and synthetase molecules need not necessarily involve a similar recognition pattern in every case, for example the interaction between one enzyme and substrate may involve a specific recognition of the amino acid stem of tRNA, whereas another interaction may involve, say, loop I.

2.6 Elongation factors

Elongation factors are enzymes which are necessary for protein synthesis, but are not an integral part of the ribosome. They were discovered when protein synthesis was studied in partially purified cell-free systems using washed ribosomes, polyU as mRNA and Phe-tRNA as amino acid donor. The enzymes are responsible for first binding amino acyl-tRNA to the A-site of the ribosome, and these are called elongation factor T (EF-T) in bacteria and elongation factor I (EF-I) in eukaryotes. The second elongation factor is needed for translocation of peptidyl-tRNA from the A-site to the P-site after peptide bond formation. This enzyme is called elongation factor G (EF-G) in bacteria and elongation factor II (EF-II) in eukaryotes.

2.6.1 Elongation factors T and I

EF-T is composed of two polypeptides called Tu and Ts (u because it is unstable, s because it is stable) and it is present in cells

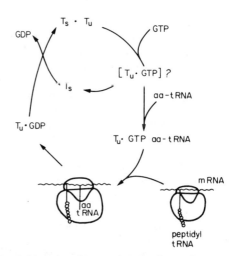

Fig. 2.39 Role of Tu and Ts in elongation (from [2]).

in large amounts (about 1 molecule per ribosome). EF-Tu has a molecular weight of 40 000 daltons and is the protein responsible for binding first GTP to amino acyl-tRNA, and next both to the A-site of the ribosome. At some stage after binding GTP is hydrolysed, and the Tu: GDP complex released. The binding constant of GDP is extremely low ($\sim 3 \times 10^{-9}$ M) and the function of Ts (molecular weight 28 000) appears to be to bind to Tu: GDP to form a Tu: Ts complex and thereby release the GDP. The complex then reacts with GTP (binding constant Tu: GTP 3×10^{-7} M) with release of Ts and the cycle repeats (Fig. 2.39).

The reaction scheme outlined above has been studied using highly purified enzymes in cell-free systems. Complexes have been detected by ribosome binding and by gel filtration on Sephadex G-100. An additional curious property of Tu which provides another useful assay, is that the protein binds to Millipore filters in the presence of GDP, but when present in the tertiary complex containing Tu: GTP:αα-tRNA it passes through such filters.

Tu interacts with all amino acyl-tRNAs except for the initiator tRNA, in which case the binding, if it occurs at all, is so weak that detection is difficult. This presumably reflects the fact that the initiator tRNA binds to ribosomes in a unique manner for use in initiation, and methionine from this species is not incorporated into internal positions of polypetides. The site on the ribosome to which EF: Tu binds is discussed later.

Factor Tu has been crystallized and it will be of interest to examine its tertiary structure as an example of another enzyme which interacts with nucleic acid and nucleotides, but to date little is known of the structure/function relationships of this molecule.

In animal cells the corresponding enzyme is EF-I; this has a minimum molecular weight of 50 000. There is no animal equivalent of Ts and this is probably related to the weaker binding of GDP to EF-I, perhaps spontaneous dissociation occurs more readily [48].

Animal elongation factor 1 occurs in several molecular weight forms ($5 \times 10^4 - 1.5 \times 10^6$) although the active species is the 50 000 monomer. The high molecular weight forms appear to be aggregates containing many copies of the enzyme complexed with lipid material particularly cholesterol esters. The binding of GTP to the high molecular weight form (EF-I_H) causes release of EF-I_L:GTP, but it is not yet known if recycling necessarily occurs via the high molecular weight form (Fig. 2.40). The occurrence of lipid bound structures containing enzymes involved in elongation is a reminder that protein synthesis in animal cells often takes place on membranes. Perhaps these complexes represent some architectural feature of cells which as yet is not well characterized.

2.6.2 Elongation factors G and II
Elongation factor G or translocase is the enzyme

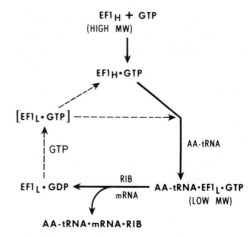

Fig. 2.40 Role of EF-I_L and EF-I_H in elongation (from [49]).

responsible for catalysing the movement of ribosomes along the mRNA. It has a molecular weight of 80 000 and is probably composed of two subunits. Its activity, like that of EF-T, is dependent on the presence of free thiol groups. Activity is measured by the change in puromycin sensitivity of bound αα-tRNA or peptidyl-tRNA, by analysis of the peptides made in response to defined polymers, or by hydrolysis of GTP. Antibodies to EF-G have also been produced and these inhibit both reactions in cell-free systems. The eukaryotic enzyme EF-II is very similar to its bacterial counterpart.

One major inconsistency in studies of protein synthesis *in vitro* is the number of GTP molecules hydrolysed per peptide bond formed. Early measurements by Lipmann in unfractionated systems indicated one GTP was required per peptide bond. However, such measurements are extremely difficult because many non-specific GTPases are present. Nevertheless, each of the two elongation factors discussed above appear to require hydrolysis of GTP for activity *in vitro* and this would indicate two GTP molecules are hydrolysed for each amino acid polymerised. Recent experiments indicate that the site of GTPase activity of both EF-T and EF-G on the ribosome is similar. Thus both GTPase activities occur on the 50S subunit; the antibiotic thiostrepton inhibits both activities (as well as initiation and termination factor GTPases); the irreversible binding of one factor (e.g. by forming an EF-G; GDP: 50S subunit complex in the presence of fusidic acid) prevents subsequent binding of the other factor and finally binding of antibodies to ribosomal proteins L_7 or L_{12} inhibits the binding of both factors. These results perhaps suggest that the GTP hydrolysis steps in protein synthesis occur at the same site, and in some concerted manner, such that *in vivo* only one GTP is hydrolysed per peptide bond. The two GTPase activities observed *in vitro* may be an artefact.

2.7 Peptidyl transferase

Peptidyl transferase, the enzyme which catalyses peptide bond formation, is a true ribosomal protein and attempts to isolate active enzyme have met with only limited success. It can be studied using isolated 50S subunits and fragments of amino-acyl tRNA. In the presence of 20% ethanol reaction is possible between, say, CAACCA fMet and puromycin. Studies with other substrates have shown that the minimum requirements *in vitro* are CCA-blocked amino acid in the P-site and amino acyl adenosine in the A-site. Besides promoting peptide bond formation, peptidyl transferase can under certain conditions also catalyse the formation of ester bonds and thus the hydrolysis of peptidyl-tRNAs leading to the release of terminated peptide chains. It is probable therefore that the same enzyme is involved in transpeptidation and in termination of protein synthesis.

2.8 Initiation factors

Many of the first studies of protein synthesis *in vitro* used synthetic polymers as mRNA.

	Elongation factor T	*Initiation factor 2*
Mol. wt.	68 000	80 000
GTP requirement	+	+
binds to	70S	30S
substrate	$\alpha\alpha$-tRNA	Met-tRNA$_F$
GTPase inhibited thiostrepton	+	+

Fig. 2.41 Comparison of EF-T and IF-2.

Efficient translation of such messenger requires abnormally high (Mg^{++}) but can utilize ribosomes washed with high concentrations of salt. When translation of natural mRNA is attempted the requirements for protein synthesis become more stringent and there is an absolute requirement for so-called initiation factors. These are protein molecules which are present in the ribosome high salt wash fraction.

Even after several years' study the detailed function of bacterial initiation factors is not yet certain, and in eukaryotes the situation is even more confusing. It is perhaps misleading to separate the functions of the various factors because they readily form complexes with one another and probably they act cooperatively; possibly for this reason the role of individual initiation factors remains uncertain. For the purposes of this discussion, however, bacterial initiation factors will be classified into three groups [49]. Initiation factor 2 is defined as that required for the binding of the initiator tRNA to ribosomes. Initiation factor 3 comprises that class of factors required for binding messenger RNA. The function of initiation factor 1 is rather ill-defined, but it may have activity in the dissociation of ribosomal subunits and in stabilizing the initiation factor complex.

Initiation factor 2 in many ways resembles EF-T (Fig. 2.41). The protein has a molecular weight of 80–100 000, and specifically sti-

	Mol. wt.	Recognizes	GTP requirement
Bacterial			
R_1	80 000	UAA; UAG	−
R_2	80 000	UAA; UGA	−
S	70 000		+
Mammalian			
TF	150 000	UAA, UAG, UGA	+

Fig. 2.42 Properties of termination factors.

mulates the binding of fMet-tRNA$_F$ to 30S subunits. In the presence of 50S subunits GTP hydrolysis occurs, apparently at the same site as EF-T and EF-G catalysed hydrolysis.

Initiation factor 3 has a molecular weight of 23 000. Protein synthesis directed by natural messenger is stimulated by the presence of this factor, but it has little effect with synthetic mRNA. This led to suggestions that it is responsible for recognizing the site on mRNA to which ribosomes bind and stimulating such binding. There is some evidence for multiple species of IF-3 and much discussion as to the possible physiological role such factors might have in determining the specificity of mRNA translation has taken place (see Section 3.3.3 p. p. 54). IF-3 also stimulates the dissociation of 70S ribosomes into subunits in preparation for subsequent initiation.

The role of IF-1 is unknown. This protein has a molecular weight of 9 000 and appears to stimulate many of the activities of IF-2 and IF-3, such as dissociation of ribosomes, recycling of factors and binding of factors to ribosomes. It is perhaps related to IF-2 in the same way as EF-Ts is related to EF-Tu, and allows recycling of the other initiation factors.

In eukaryotes even more initiation factors are known. A GTP-dependent IF-E2 is well characterized, it catalyses the binding of initiator tRNA. Messenger binding factors (IF-E3) some with mRNA specificity, are also known, but in addition other factors, one having ATPase activity have also been reported. Their exact function is unknown.

2.9 Termination factors

Genetic experiments in which the amino acid replacements present in revertants of chain termination mutants were analysed indicated that three chain termination codons are present in bacteria, viz.: UGA, UAA and UAG. Experiments using mutant bacteriophage mRNA with termination codons present at known positions in the mRNA showed that termination is an active process requiring not only a termination codon, but also several protein factors. There is no requirement for a terminator tRNA. The detailed study of this process was facilitated by a simple triplet-dependent assay *in vitro* [50]. Initiator tRNA was first bound to the ribosome, and the requirements for release of fMet in the presence of termination codons studied. In this way two release factors were detected, R_1 which reads UAA and UAG, and R_2 which reads UAA and UGA; both have a molecular weight of about 50 000. A third stimulatory factor, molecular weight 70 000 (called S or R_3) is also required; this binds GTP and GDP but a clear requirement for GTP hydrolysis has not been shown in bacteria. It seems likely however that GTP hydrolysis is involved and that hydrolysis takes place at the same ribosomal site as EF-T, EF-G and IF-2 dependent GTP hydrolysis, because thiostreption, antibody to L_7 and L_{12}, and prior binding of EF:G:GDP: fusidic acid all inhibit termination reactions *in vitro*. In mammalian systems only one termination factor with a molecular weight of 150 000 has been characterized. It shows a clear GTPase activity and recognizes all three termination codons.

References

[1] Jacob, F. and Monod, J. (1961), *J. Mol. Biol.*, **3**, 318-356.
[2] Brenner, S., Jacob, F. and Meselson, M. (1961), *Nature*, **190**, 576-580.
[3] Nirenberg, M. and Matthaei, J. (1961), *Proc. Nat. Acad. Sci. (USA)*, **47**, 1588-1602.
[4] Lukanidin, E.M., Georgiev, G.P. and Williamson, R. (1971), *FEBS Letters*, **19**, 152
[5] Aviv, H. and Leder, P. (1972), *Proc. Nat. Acad. Sci. (USA).*, **69**, 1408-1412.
[6] Mathews, M.B. (1973), in *Essays in Biochemistry*, (ed.) Campbell, P.H. and Dickens, F. Biochemical Society, Academic Press, London and New York, pp. 59-102.
[7] Mathews, M.B. (1972), *Biochim. Biophys. Acta*, **272**, 108-118.
[8] Roberts, B.E. and Paterson, B.M. (1973), *Proc. Nat. Acad. Sci. (USA)*, **70**, 2330-2334.
[9] Smith, A.E., Wheeler, T., Glanville, N. and Kaariainen, L. (1974), *Europ. J. Biochem.*, **49**, 101-110.
[10] Gurdon, J.B., Lane, C.D., Woodland, H.R. and Marbaix, G. (1971), *Nature*, London, **223**, 177.
[11] Gould, H.J. and Hamlyn, P.H. (1973), *FEBS Letters*, **30**, 301-304.
[12] Brownlee, G.G. (1972), *Determination of Sequences in RNA*, North-Holland, Amsterdam.
[13] Steitz, J.A. (1969), *Nature*, London, **224**, 957-964.
[14] Shine, J. and Dalgarno, L. (1975), *Nature*, London, **254**, 34-38.
[15] Jeppesen, P.G.N., Steitz, J.A., Gesteland, R.F. and Spahr, P.F. (1970), *Nature*, London, **226**, 230-237.
[16] Wachter, R.D., Merregaert, J., Vandenberghe, A., Contreras, R. and Fiers, W. (1971), *Europ. J. Biochem.*, **22**, 400-414.
[17] Sanger, F. (1971), *Biochem. J.*, **124**, 833.
[18] Min Jou, W., Haegemann, G., Tsebaert, M. and Fiers, W. (1972), *Nature*, London, **237**, 82-88.
[19] Darnell, J.E., Jelinek, W.R. and Molloy, G.R. (1973), *Science*, **181**, 1215-1221.
[20] Sheiness, D., Puckett, L. and Darnell, J.E. (1975), *Proc. Nat. Acad. Sci. (USA).*, **72**, 1077-1081.
[21] Adams, J.M. and Cory, S. (1975), *Nature*, London, **255**, 28-33.
[22] Gielkens, A.L.J., Berns, T.J.M. and Bloemendal, H. (1971), *Europ. J. Biochem.*, **22**, 478-484.
[23] Davis, B.D. (1971), *Nature*, London, **231**, 153-157.
[24] Leder, P. and Nirenberg, M. (1964), *Science*, **145**, 1399.
[25] Monro, R.E. and Marcker, K.A. (1967), *J. Mol. Biol.* **25**, 347.
[26] Monro, R.E. (1967), *J. Mol. Biol.*, **26**, 147-151.
[27] Erbe, R.W., Nau, M.M. and Leder, P. (1969), *J. Mol. Biol.*, **38**, 441-460.
[28] Monier, R. (1972), in *The Mechanism of Protein Synthesis and its Regulation*, (ed.) Bosch, L., North-Holland, Amsterdam, pp.353.
[29] Richter, D., Erdmann, V.A. and Sprinzl, M. (1973), *Nature New Biol.*, **246**, 132.
[30] Barrel, B.G. and Clark, B.F.C. (1974), *Handbook of Nucleic Acid Sequences*, Joynson-Bruvvers Ltd., Oxford.
[31] Wittmann, H. (1972), in *FEBS Symposium*, Vol. 23, (ed.) Cox, R.A. and Hadjiolov, A.A., Academic Press, London, pp. 1-17.
[32] Kurland, C.G., Donner, D., Van Duin, J., Green, M., Lutter, L., Randall-Hazelbauer, L., Schaup, H.W. and Zeichhardt, H. (1972), *FEBS Symposium* **27**, pp. 225.
[33] Stoffler, G. and Wittmann, (1972), in *The Mechanism of Protein Synthesis and its Regulation*, (ed.), Bosch, L., North-Holland, Amsterdam, pp.285.
[34] Bode, U., Lutter, L.C., Kurland, C.G., Zeichardt, H. and Stoffler, G. (1974), *Acta biol. med. germ.*, **33**, 625-628.
[35] Mizushima, S. and Nomura, M. (1970), *Nature*, London, **226**, 1214.
[36] Cannon, M. and Cundliffe, E. (1973), in *Techniques in Protein Biosynthesis*, Vol. 3 (ed.) Campbell, P.N. and Sargent, J.R., Academic Press, London.
[37] Nomura, M. (1973), *Science*, **179**, 864.
[38] Crick, F.H.C. (1966), *J. Mol. Biol.*, **19**, 548.
[39] Chapeville, F., Lipmann, F., von Ehrenstein, G., Weisblum, R., Ray, W.J. and Benzer, S. (1962), *Proc. Nat. Acad. Sci. (USA)*, **48**, 1086-1092.
[40] Piper, P.W. and Clark, B.F.C. (1974), *Europ. J. Biochem.*, **45**, 589-600.
[41] Robertus, J.D., Ladner, J.E., Finch, J.T., Rhodes, D., Brown, R.S., Clark, B.F.C. and Klug, A. (1974), *Nucleic Acids Research*, **1**, 927-932.
[42] Robertus, J.D., Ladner, J.E., Finch, J.T., Rhodes, D., Brown, R.S., Clark, B.F.C. and Klug, A. (1974), *Nature*, London, **250**, 546.
[43] Smith, J.D. (1972), *Ann. Rev. Gen.*, **6**, 235.
[44] Rudland, P.S. and Clark, B.F.C. (1972), in *The Mechanism of Protein Biosynthesis and its Regulation*, (ed.), Bosch, L., North-Holland, Amsterdam, pp.55.
[45] Smith, A.E. and Marcker, K.A. (1970), *Nature*, London, **226**, 607-610.
[46] Piper, P.W. (1975), *Europ. J. Biochem.*, **51**, 283.
[47] Bruton, C. and Hartley, B.S. (1970), *J. Mol. Biol.*, **52**, 165.
[48] Weissbach, H. and Brot. N. (1974), *Cell*, **2**, 137.
[49] Revel, M., Pollack, Y., Groner, Y., Scheps, R., Monye, H., Berissi, H. and Zeller, H. (1972), in *FEBS Symposium*, **27**, pp.261.
[50] Beaudet, A.L. and Caskey, C.T. (1972), in *The Mechanism of Protein Synthesis and its Regulation*, (ed.) Bosch, L., North-Holland Amsterdam, pp. 133.

3 The mechanism of protein biosynthesis and its control

3.1 Introduction
In the last chapter the structure of many of the components of protein synthesis was described in some detail. In this chapter the dynamics of synthesis are considered and the involvement and interdependence of the molecules described above will be emphasized. In addition, the control of protein synthesis will be mentioned and the possible molecular mechanisms of such control discussed.

It should be emphasized from the beginning that the major control on the expression of genetic information appears to act at the level of transcription of DNA into mRNA. Thus the synthesis of a protein is mainly dependent on the presence of the mRNA coding for it. Nevertheless there are instances where preformed mRNA is not translated and other instances where additional fine control at the translational level occurs.

Translational control can be thought of as operating whenever the amount of a given message translated does not simply reflect the amount of messenger available for translation. Within this definition two basic types of regulation are possible, general, quantitative control and selective, qualitative control. Clearly the molecular mechanisms involved would be expected to differ in the two instances. Quantitative control can be affected by changing the synthetic pathway such that it becomes rate-limiting, whereas qualitative control mechanisms must somehow distinguish between the synthesis of two different proteins and select-ively affect only one of them. Presumably some distinguishing features must be encoded within mRNA itself to allow such discrimination.

In addition to the cytoplasmic regulatory mechanisms that apply to the synthesis of all proteins, translational control events can be classified into two groups; those occurring as a result of (1) changes in cellular conditions, for instance during growth control (e.g. by amino acid or serum depletion) and during the cell cycle (e.g. mitosis); and (2) changes in the physiological state of the organism, for example during fertilization and early development, during hormonal regulation and during phage or viral infections.

The structure of the various components of protein synthesis machinery will itself have a profound effect on the translation of a given mRNA, but in addition a whole battery of physiological effectors can also influence translation. These effectors may be primary signals in themselves or secondary products of some other stimulus. Some examples of biochemicals known to be active in this way are dsRNA, haemin, cAMP, hormones, antibiotics, drugs, toxins, colicins and interferons. In the following sections the possible site of action of such biochemicals and their occurrence in relation to the regulatory mechanisms mentioned above will be described.

3.2 mRNA metabolism
In bacteria, provided that RNA polymerase molecules are able to attach to the promoter of any particular gene, mRNA synthesis takes place

and ribosomes attach to the nascent mRNA chain very soon after initiation of RNA synthesis. There appears to be little RNA processing and transport and very little likelihood of any control mechanism operating at this level.

In eukaryotic cells on the other hand, a large number of events occur after synthesis of mRNA sequences and before the mRNA eventually attaches to a ribosome. These include cleavage and breakdown of large sections of the primary transcript, (HnRNA) polyadenylation and modification of the nascent transcript, the attachment of various proteins (first in the nucleus and secondly in the cytoplasm) and transport of the mRNA from nucleus to the cytoplasm. It must be admitted that the exact details of these various events are still quite unclear and this reflects the difficulty in studying the metabolism of a population of molecules of very high molecular weight. HnRNA in many cases appears to have a molecular weight of several million daltons and to study cleavages, modification of minor bases, polyadenylation etc. of one particular species of HnRNA is as yet impossible. The study of the synthesis of viral mRNA from integrated DNA may prove a useful model system in this respect. Nevertheless it is too soon to state with confidence whether one or more different pathways for the synthesis and metabolism of mRNA exist in animal cells or if any specific control mechanisms operate after transcription but before translation. For the purposes of this discussion it will be assumed that mRNA somehow arises and finds its way to the cytoplasm. The discussion of translation and its control will be restricted here to events occurring in the cytoplasm. Fig. 3.1 summarizes the current overall model for protein synthesis that will be used as the basis for discussion.

3.3 Initiation complex formation

The initiation of protein synthesis is probably the most critical step in protein biosynthesis, for provided that this step is successfully achieved, elongation of the polypeptide chain continues with little further control. The initiation step is a complex reaction involving many cellular components and this is emphasized by the extreme sensitivity of initiation to a wide variety of agents. Initiation in bacteria is more sensitive to low temperature than elongation, and in eukaryotes it is also very sensitive to elevated temperatures. Many antibiotics act at initiation (e.g. pactamycin, ediene), as do many drugs (e.g. aurin tricarboxylic acid (ATA), sodium fluoride); biochemicals (e.g. dsRNA, haemin) and even elevated levels of salt! Although the exact mechanism of action of all of these is not known they are of considerable value in studying protein synthesis, and an understanding of their activity will ultimately be of great value in understanding initiation. The initiation process can be divided into two separate stages, first the binding of the initiator tRNA and second the binding of mRNA.

3.3.1 Binding of initiator tRNA

Until recently it was thought that mRNA bound first to the small ribosomal subunit followed by a codon-dependent binding of the initiator tRNA. It has very recently been shown that this is probably not the correct order and that in fact the initiator tRNA binds to free 30S or 40S subunits in both bacterial and mammalian systems. This was first shown by incubating reticulocyte lysates with purified initiator tRNA and demonstrating on sucrose gradients that a large percentage of subunits had labelled tRNA attached but not mRNA [1]. Subsequently it has been shown that the binding is absolutely dependent on the initiation factor IF2 and GTP. The corresponding complex in bacteria (which sediments at 34S as opposed to the free subunit at 30S) is quite unstable and in both cases the bound fMet-tRNA is freely exchangeable with added initiator.

In the past there has been considerable de-

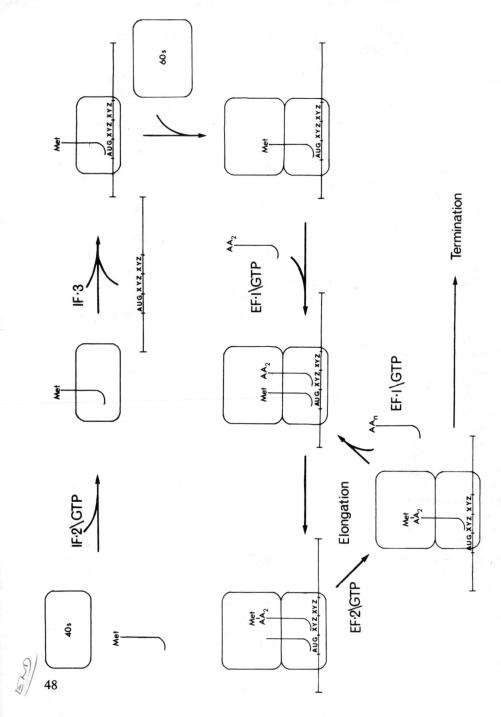

Fig. 3.1 General model of protein synthesis.

bate as to whether the bacterial initiator tRNA binds, like all other $\alpha\alpha$-tRNAs, to the A-site of the ribosome or to the P-site [2]. From the beginning it was clear that the binding was unusual and the most compelling evidence that the P-site was involved was the finding that bound fMet-tRNA could react with puromycin to give fMet-puromycin, and that binding of initiator was partially insensitive to antibiotics such as tetracycline that inhibit binding at the A-site. The position was complicated by the finding that GTP hydrolysis was required before fMet-tRNA could react with puromycin. Thus, if the GTP analogue GMPPCP (which cannot be hydrolysed by GTPases to give GDP, and Pi, with release of energy) was used, the bound initiator was in some way inactive. The activation, on GTP hydrolysis, was shown not to be dependent on the enzyme EF-G as in the translocation reaction. With the finding that initiator tRNA binds first to ribosomal subunits before addition of mRNA it became clear that the initiator tRNA binding site could not be identical to the P-site (as defined in Section 2.2.3), but that on addition of mRNA and 50S subunits it is eventually positioned in the P-site without going through the A-site. How this occurs is not yet completely clear, but there is as yet no direct evidence to support a model for initiation proposed by Bretscher in which the 50S subunit attached to the fMet-tRNA: 30S complex to form a hybrid site for the initiator composed of the 30S P-site and the 50S A-site [3]. The finding that eukaryotic initiator tRNA does not have the sequence GTΨCG in loop IV, which is present in almost all other tRNAs and is thought to be involved in binding tRNA to ribosomes, is also consistent with the idea that it binds at another ribosomal site.

Recent experiments in which the identity of the ribosomal proteins and RNA composing the various sites on the ribosome are yielding some information (see Section 2.3.5). For instance some proteins which will react with chemical affinity labelled initiator tRNA in 70S initiation complexes and thus identify the P-site, also react with labelled tRNA bound in the A-site. Similarly the GTPase activity of IF-2 is inhibited by agents that inhibit the other ribosome-associated GTPases of elongation and termination factors. Perhaps the different ribosomal sites are much more closely related than had previously been thought. This further emphasizes the fact that to date analysis of the kind just mentioned is still not sufficiently advanced to characterize in detail the different molecular interactions involved in the various intermediate initiation complexes and instead only looks at stable readily-formed complexes (e.g. the fMet-tRNA: R_{17} mRNA:70S complex). Dynamic intermediates must still be identified by less rigorous, but more sensitive techniques such as those involving antibiotic sensitivity, GTP hydrolysis and sucrose gradient sedimentation.

The binding of initiator tRNA in eukaryotes occurs via a similar mechanism to that described for bacteria with the added experimental advantage that some of the intermediates (for example IF_2:GTP:Met-tRNA$_F$ and Met-tRNA: 40S subunit) are more stable, and therefore more readily analysed. Although as yet no details are known of the ribosomal proteins involved there is increasing evidence to suggest that the amount of subunit bound Met-tRNA$_F$ is a significant control of the level of protein synthesis in animal cells.

The amount of globin synthesized in intact reticulocytes or in lysates is dependent on added haemin. In its absence a pro-inhibitor is converted to its active form and protein synthesis is rapidly reduced. The details of the inhibition are not yet clear, but parallel with the reduced protein synthetic activity is a reduction in the amount of subunit-bound Met-tRNA$_F$ [4]. Similarly, globin synthesis is inhibited by extremely low concentrations of dsRNA; again the reduced synthetic activity is accompanied

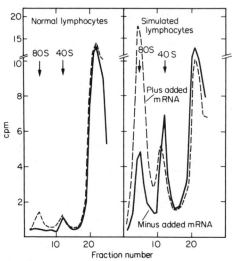

Fig. 3.2 Initiation complex formation in normal and PHA-stimulated extracts from lymphocytes (from [5]).

by a reduction in bound initiator tRNA.

This phenomenun may be more general, for when cultured cells are transferred from normal to depleted medium the level of protein synthesis is rapidly reduced, and decreased levels of Met-tRNA:40S subunit complex are found. When the starved cells (e.g. depleted of an amino acid or serum) are returned to complete medium or when lymphocytes are stimulated to synthesize greater amounts of protein by phytohaemagglutinin, the levels of complex rise (Fig. 3.2). It is not obvious how changes in the amount of initiation complex could cause a drastic qualitative change in protein synthesis; nevertheless it is a most effective quantitative control. There is as yet no clue as to the lesion involved in cells containing reduced levels of bound initiator tRNA; it could be a change in the charging of initiator tRNA, the activity of IF-2 or the stability of the complex once formed.

3.3.2 Binding of messenger RNA

The binding of messenger to the initiator-tRNA: small ribosomal subunit initiation complex is probably one of the most critical steps in protein biosynthesis. It is a step which can be controlled quantitatively, but more important it could be the major site of qualitative control of protein synthesis in the cytoplasm. For these reasons it has been the subject of intense study, but even today the process is not completely understood. For the purpose of this discussion the formation of the initiation intermediate containing mRNA will be considered in terms of the mRNA itself, the ribosome and initiation factors.

mRNA. An early experiment by Dintzis, in which the amount of label in each of the tryptic peptides of completed globin was measured, following a short radioactive pulse, clearly showed that globin is made from the end with a free amino terminal (which was only lightly labelled) to the end with a free carboxyl group (which contained most radioactivity) [6]. Other experiments, mainly with synthetic polymers of defined sequence, established that the direction of reading the mRNA was from the end with a free 5'OH to the end with a free 3'OH. That the ribosome does not merely attach to the 5'-end of mRNA was demonstrated by showing that the circular DNA from phage fd can be translated *in vitro* in the presence of neomycin, to give accurate synthesis of at least one phage protein. Together these experiments suggested that at the initiation of protein synthesis ribosomes attached to some site, which itself is not necessarily translated, toward the 5'-end of the coding sequence in mRNA. Such a model has now been firmly established following the complete determination of the coat protein gene of phage MS2, including the non-translated regions (see Section 2.1.3). What is not so clear, however, is what dictates the position to which ribosomes attach.

Since there is only one methionine codon, but at least two species of tRNA^{Met} with very different properties, it is clear that some structural feature of mRNA must indicate that some AUG codons in mRNA signify initiation and are decoded by initiator tRNA, whereas others indicate internal methionine and are read by Met-tRNA$_M$. Other AUG sequences in mRNA are out of phase and are never read as methionine. The structural feature differentiating the initiating AUG could be either the primary sequence or the overall structure.

Analysis of the nucleotide sequence of ribosome binding sites from several mRNAs (mostly from the RNA phages) was discussed earlier (Section 2.1.3) and suggests three important points. First, some common sequences appear in the primary sequence of mRNA in the untranslated region immediately adjacent to the 5' side of the initiating AUG residue. This is discussed again below (Section 3.3.3 p. 53). Secondly, some ribosome binding sites (e.g. RNA phage coat protein site) can be drawn as hairpin-like structures, with AUG in an exposed position, whereas many internal AUG sequences are involved in secondary structure and are not exposed. Finally other binding sites (e.g. RNA phage replicase site) are not exposed and these appear to be partially covered or closed by interactions with other parts of the mRNA molecule.

The importance of the AUG codon in phasing mRNA read-out is indicated by experiments in which synthetic polymers with AUG positioned at different distances (1–10 nucleotides) from the 5'-end were translated *in vitro*. Each mRNA directed the incorporation of methionine from Met-tRNA$_F$, even if the AUG was preceded by up to three codons. For example $(AAA)_3 AUG(U)_{35}$, which might otherwise have been expected to make the peptide $(Lys)_3$. Met. $(Phe)_{10}$ with the methionine residue donated by Met-tRNA$_M$, in fact directed synthesis of Met-$(Phe)_{10}$ with methionine from the initiator tRNA [7].

The recently described modified bases which appear to be present at the 5' terminus of many

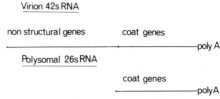

Fig. 3.3 Semliki Forest virus mRNAs.

eukaryotic mRNAs may be important in the interaction between ribosomes and mRNA. As yet there is no direct evidence to support this. Poly-A and the proteins which are found attached to almost all mRNA species may also be involved in initiation complex formation, but again there is little direct evidence for this. Recent results suggest that during animal virus infection, viral proteins may become specifically attached to viral mRNA, thus raising the interesting possibility that such proteins play a role in deciding which viral mRNAs are translated, either by covering up the initiation site and preventing translation (cf. Qβ replicase closing the coat protein initiation site, see Section 3.6) or favouring the interaction between mRNA and ribosome (i.e. behaving as an initiation factor).

The animal virus mRNAs also illustrate other interesting cases of translational control occurring at the level of initiation complex formation [8]. Reovirus contains ten segments of double-stranded genomic RNA, each of which is transcribed to give an mRNA species. Although infected cells contain approximately equal amounts of the different size classes of viral mRNA, very different amounts of the corresponding proteins are made. This suggests that the rate of initiation differs for different mRNAs, and this probably reflects differences around the ribosome binding sites, in each mRNA. Arboviruses such as Semliki Forest virus contain plus strand single-stranded RNA as the genomic material. Virion RNA has two sites for the initiation of protein synthesis but one of these (for the coat protein cistron) is very inefficient compared with the other (for the non-structural protein cistron). However, infected cells contain a polysomal 26S mRNA which codes for the coat proteins and is a very efficient messenger (Fig. 3.3). Thus the same nucleotide sequence (i.e. in 42S or 26S RNA) can behave differently, depending on the structure of adjacent RNA. Presumably the structure of the virion RNA makes the coat protein initiation site inaccessible, whereas in 26S RNA the site is opened up (cf. the replicase binding site of RNA bacteriophage RNA, Section 2.1.3 and 3.6).

The role of mRNA in formation of the initiation complex may be summarized thus: sequence analysis of mRNA (admittedly mainly from phage mRNAs which may not be typical), indicates that ribosomes attach to a position on the messenger to the 5′ side of the coding sequence, at a site which is characterized by at least part of a common primary sequence and to which binding is influenced by the accessibility of the site as determined by its secondary and tertiary structure, or its interaction with other molecules. The binding of mRNA to the ribosome also involves the interaction of a specific AUG sequence in the mRNA with prebound initiator tRNA. Such binding however is also influenced by the ribosome itself and initiation factors, and this is discussed in the next sections.

Ribosomes. At one time, it seemed likely that initiation factors were able to recognize the structure of the initiation site region of mRNA and, depending on the stability of the complex formed between them, the binding of mRNA to the ribosome varied. However, it is now known that the ribosome plays an active role in mRNA selection. The three cistrons of RNA phage R_{17} are translated in *E. coli* cell-free systems in approximately the ratio found in infected cells (say 20 coat protein: 5 replicase: 1 A-protein). If the same mRNA is translated

E. coli
16S rRNA HO A U U C C U C C A C Py

A protein (5′)	C U A G G A G G U U U
Replicase	A C A U G A G G A U U
Coat	A C C G G G G U U U G

B. stearothermophilus
HO A U C U U U C C U Py

A protein (5′)	U U C C U A G G A G (3′)
Replicase	A C A U G A G G A U
Coat	A A C C G G G G U U

Fig. 3.4 R_{17} ribosome binding site sequences and proposed pairing with the 3′ terminus of 16S RNA (from [10]).

in *Baccilus stearothermophilus* extracts only the A-protein is synthesized. Surprisingly the lesion lies not in the initiation factor fraction, but in the 30S ribosomal subunit. Thus reconstituted cell-free systems containing *E. coli* 30S subunits plus all the other components from *B. stearothermophilus* make all three proteins [9]. Further reconstitution experiments have shown that the *E. coli* components responsible are the 16S RNA and ribosomal protein S12.

The isolated ribosome binding sites of R_{17} are themselves able to bind to ribosomes, but the ratio of the fragments bound (viz., 40:5:1; A-site: replicase: coat) is quite different from the amount of protein synthesized from each site in total RNA. Whole phage RNA previously treated with formaldehyde, or heated such that secondary structure is disrupted is also able to bind to ribosomes. It directs the synthesis of the three phage proteins in quite different ratios to native RNA and also makes many non-physiological polypeptides.

These experiments all show that ribosomes have an intrinsic ability to select some initiation sites on mRNA. The sequence of the 3′OH terminus of 16S RNA from several species of bacteria indicates that the intrinsic capacity of a given species of ribosome may be directly related to the degree of complementarity between the 16S rRNA and the common sequence in the ribosome binding site (GGAGGU) (Fig. 3.4). Thus the 3'-end of *E.coli* 16S and R_{17} RNA can form 7, 4 and 3 base pairs respectively with the ribosome binding sites of R_{17} A: replicase: and coat cistrons, whereas for *B. stearothermophilus* the corresponding numbers are 4: 4: 2. Since the replicase binding site is inaccessible in whole RNA because of secondary and tertiary structure (Section 2.1.3), this result fits well with the experimental data, viz. *B. stearothermophilus* ribosomes initiate only at the A-site. Two base pairs are considered insufficient for binding at the coat protein site, and if ribosomes do not attach at this position and move along the mRNA the replicase site remains closed. The fact that in *E.coli* under physiological conditions only small amounts of A-protein are made suggests that the A-protein initiation site is closed by secondary structure. Presumably, since the cell-free system from *B. stearothermophilus* is incubated at about 45°C rather than 37°C for *E.coli* the secondary structure is disrupted and the binding site accessible.

Ribosomal proteins S12 and S1 which bind near to the 3'-end of 16S RNA, are also involved in this interaction, and these are mentioned later. The importance of the 3'-end of 16S RNA is further indicated by the loss of activity of ribosomes from which this portion has been cleaved by Colicin E_3. Although the relationship between complementarity and amount of isolated fragment bound to ribosomes holds well it is not consistent with the total amount of protein synthesized on unfragmented mRNA. Other factors must also be involved, and these are discussed below.

The RNA:RNA interactions just described, suggest a rational explanation for translational control of different mRNAs, in that the efficiency of initiation can be predicted in part by understanding the interaction between rRNA and mRNA. Such a mechanism can control the efficiency with which any given mRNA and ribosome will interact; however, it cannot be invoked to explain changes in protein synthesis within the same cell using the same population of ribosomes, since the structures of the mRNA and rRNA are invariant. However, such qualitative control can be accommodated by assuming that ribosomal proteins influence the interaction between the RNA species and that the activity of the proteins may be changed by say, phosphorylation. Thus the switching on of the synthesis of a protein on a pre-formed, but previously inactive mRNA could occur by such a mechanism, as in the rapid turn-on of histone synthesis using maternal mRNA after fertilisation. Similar changes in ribosomal proteins could be involved in translation control during phage and virus infection, particularly since it is known that many virions contain kinases.

Initiation factor IF-3. Initiation factor 3 is required for the translation of natural mRNA but not of synthetic polymers. IF-3 has some activity in the dissociation of 70S ribosomes into subunits and some of its properties could merely reflect this ability, for example synthetic mRNAs can bind directly to 70S ribosomes whereas natural mRNAs must go through subunit intermediates. Nevertheless it seems clear that IF-3 does have a second, specific role in the binding of mRNA to ribosomes and that it interacts with mRNA either directly or once it is bound to the ribosome. One possibility is that mRNA binds weakly to ribosomes as directed by its complementarity with rRNA and that IF-3 stabilizes such binding. In much the same way $\alpha\alpha$-tRNA can bind in a codon-dependent non-enzymatic reaction, but such binding is greatly stabilized by elongation

Assay		Initiation Factor IF-3 fraction			
		B1	B2	B3	B4
(a) amino acid incorporation ratio of activity	$\frac{T_4 \text{ mRNA}}{MS2 \text{ RNA}}$	1·9	1·3	3·8	31·0
(b) formyl methionine incorporation ratio incorporated into	$\frac{\text{coat}}{\text{synthetase}}$	0·6	3·0	1·0	
(c) ribosome binding to MS2 RNA ratio binding to	$\frac{\text{coat site}}{\text{synthetase site}}$	0·75	2·5	1·4	

Fig. 3.5 Activity of different fractions of E.coli IF-3 (from [11]).

Factor	mRNA	% ribosome bound
none	globin	4
reticulocyte	globin	25
muscle	globin	13
none	myosin	5
reticulocyte	myosin	7
muscle	myosin	33

Fig. 3.6 Specificity of initiation factors in binding mRNA to reticulocyte ribosomes (from [13]).

factors.

Although the exact mechanism of action of IF-3 has not been established, the greatest interest in this factor concerns its ability to distinguish between different mRNA species and thereby cause selective translation of different mRNAs [11]. Several reports claim that IF-3 from E.coli can be separated into at least two activities on chromatography of highly purified factor. The function of the separated IF-3 fractions can be tested by their ability to form known proteins in response to specific mRNAs in highly fractionated, IF-3-dependent cell-free systems, to form initiation complexes containing labelled fMet-RNA using different mRNA species, or by RNA fingerprint analysis of the ribosome binding sites to which ribosomes attach under the influence of the different factors (Fig. 3.5). Thus E.coli IF-3 fractions differ in their ability to translate mRNA made either early or late after infection with phage T4, and their ability to recognize the different ribosome binding sites on RNA phage. The experiments just described utilized IF-3 from uninfected cells, other data suggests that after infection by phage T4 a new species of IF-3 activity can be detected. Similarly there have been claims that the switch from the early to late phase of infection by phage T7 is accompanied by the appearance of a new initiation factor. There has been considerable debate as to the physiological significance of such results and as yet the ability of IF-3 factors to discriminate between different messengers is not completely accepted.

The situation in bacteria was further confused by the presence of yet other factors called interference or i-factors. i-factor selectively inhibits translation from some initiation sites, for example the MS_2 coat cistron. It is now known that i-factor is identical to ribosomal protein S1, which is one of the so-called fractional proteins. Chemical studies show that S1 probably attaches near the 3'-end of 16S rRNA, and that cross-linked complexes containing proteins S1, S12 and IF-3 can be formed. It therefore seems reasonable to suppose that any ribosome binding site in a messenger attaches to the 3'-end of 16S RNA (which presumably is positioned near to the P-site region of the 30S subunit), and that protein S12 is required to keep the rRNA accessible. Protein S1 also binds near this site and excess S1 sometimes competes with mRNAs for the site. Initiation factor-3 recognizes any such complex once it is formed and greatly stabilises favoured sites whereas it has little effect on others.

Initiation factors that are required for the binding of mRNA have been described in animal systems. At least two activities appear to be involved, one referred to as IF-E3 has a molecular weight of about 500 000 daltons and contains a number of polypeptides. It is required for the translation of natural mRNA but displays no selectivity between different mRNA species. A second factor has been purified to homogeneity from the cytoplasm of mouse cells. This factor (IF_{EMC}) is specifically required for initiation on some mRNAs (e.g. EMC viral RNA) whereas it has little effect on the translation of others (e.g. globin mRNA) [12]. Similar factors have been partially purified from ribosomes from different cell types and their activity tested with homologous and heterologous mRNAs. Such experiments show for instance that reticulocyte ribosomes can make myosin but only in the presence of muscle initiation factors (Fig. 3.6). The presence of such messenger-specific factors suggests that cells contain a number of such proteins, each responsible for initiation on a different group or class of messenger.

It seems likely that cells contain a number of mRNAs coding for proteins which are needed constantly to maintain basic cellular metabolism, the so-called household proteins; their synthesis is probably subject to little translational control. In addition to these the cell can synthesize other proteins, as and when required, depending on the growth rate, physiological state, degree of differentiation etc., the so-called luxury proteins. Perhaps synthesis of the latter group is sometimes dependent on the presence of the relevant initiation factor required by the particular group or class of mRNAs involved.

The presence or absence of such molecules may also form the basis of permissivity of cells to infection by animal viruses. The lack of a required initiation factor could result in abortive infection. The best example of such control has been described in monkey cells infected by adenovirus. The virus can enter normal monkey cells, but although mRNA synthesis occurs as usual, little viral protein synthesis can be detected and no progeny virus is released. However, prior transformation of the monkey cells with SV40 makes the cells permissive, and virus production occurs normally. The change in the monkey cells after transformation has been located in the initiation factor fraction and suggests that SV40 induces some activity that modifies monkey IF-3 type factor, such that it becomes able to recognize adenovirus mRNA [14]. The molecular basis of the modification in unknown but, as in other cases, phosphorylation, methylation, glycosylation or proteolytic cleavage could be involved. It is equally possible that rather than a modification of a pre-existing factor a completely new initiation factor activity could be induced.

The initiation factors described above are all protein molecules. Recent evidence suggests that RNA species may also be required at initiation in animal systems. Thus treatment of cells with antibiotics or drugs which inhibit RNA synthesis also cause a decrease in protein synthesis. The half-life of the effect is about 90 minutes, and initially this was taken as a measure of the half-life of mRNA. More recent experiments utilizing new methods of isolating mRNA based on its poly-A content indicate that its half-life is in fact much greater (of the order of 24 hours). This led to the suggestion that another species of rapidly metabolized RNA is required for protein synthesis in addition to mRNA, rRNA and tRNA. Further experiments indicate that a low molecular weight (10 000) species of RNA is present in the initiation factor fraction and there have been claims that any one species of such RNA, referred to as translational control RNA (tcRNA), is required for initiation of protein synthesis on one particular group of mRNAs, and discriminates against translation of others. This idea was extended to include a novel mechanism for translational control, in which HnRNA contains not only mRNA sequences but also additional RNA stretches (tcRNA) that are exported with the nascent mRNA and directly affect translation of that mRNA [15].

It should be emphasized that although such a model is very elegant there is virtually no evidence to support it!

In the previous sections the initiation step of protein synthesis has been described in considerable detail and it is clear that the process is very complicated and is far from completely understood. Nevertheless the formation of the initiation complex on any one mRNA in any particular cell can be described with some accuracy in terms of the ribosome, its proteins and RNA; the mRNA and its primary, secondary and tertiary structure and the complement of initiation factors, both protein and RNA. What is less well defined are events following some physiological change in the cellular conditions, such as some hormonal signal or viral infection, and their role in qualitative control of translation. Whilst it is easy to postulate elegant models of phosphorylation of ribosomal proteins or initiation factors very little direct evidence for such events happening *in vivo* has been collected. Indeed, many authors have suggested that messenger specific translational control is ruled out by the finding that many different mRNAs are translated accurately in heterologous cell-free systems and in frog oocytes. This argument can be countered by noting the extreme inefficiency of most cell-free systems. As has been mentioned above, all mRNAs probably have an intrinsic ability to bind to ribosomes at the correct initiation site and it is hardly surprising therefore that a low level of accurate synthesis is possible in many cell-free systems. The frog oocyte is more efficient at mRNA translation but in this case the oocyte can hardly be described as a typical cell, and it may well be expected to contain a wide variety of initiation factors necessary during the early stages of development when few new ribosomes are made. Similarly most mRNAs isolated to date cannot be called typical; in the main they come from highly differentiated cells such as reticulocytes, myoblasts etc., or from viruses.

These may be peculiar in their needs at initiation. The synthesis of specific well defined proteins must be studied in a variety of highly purified cell-free systems from normal and physiologically modified cells before the role of mRNA specific initiation factors etc. can be firmly established one way or the other.

However, it is only fair to mention that there have been suggestions that virtually all changes in translational control are passive and merely result from the different stabilities or binding constants of the mRNA:ribosome complex formed at initiation [16]. Qualitative changes in protein synthesis on treatment with antibiotics or drugs, or on infection with virus could simply reflect a general overall slowing down of protein synthesis at the level of initiator tRNA binding which would in turn affect different mRNAs unequally depending on the stability of the initiation complex. Thus a quantitative change in the rate of total protein synthesis in reticulocytes alters the ratio of α to β globin chains made, presumably because the ribosome binding sites of the mRNAs differ.

3.4 Elongation

After the critical step during which the initiation complex is formed, chain elongation continues in cyclic fashion in what appears to be a fairly straight-forward manner; addition of amino acyl-tRNA, peptide bond formation, translocation, addition of amino acyl-tRNA etc. (Fig. 3.1). The enzymes involved at this stage and the ribosomal sites to which these attach are reasonably well established and have been discussed in detail in earlier sections.

The overall rate of protein synthesis in cells could be quantitatively controlled at the elongation step if it were made rate limiting because of a lack of elongation factors. Such enzymes are normally present in cells in excess with about one copy of each per ribosome; however, there are some conditions, such as starved

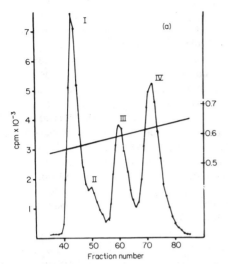 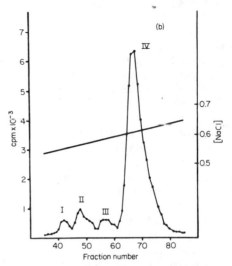

Fig. 3.7 Met-tRNAs from chick embryo cells (a) and avian myeloblastosis virus (b) from [8].

animals, where the activity of elongation factors is reported to be low and possibly rate limiting. Similarly the overall rate of protein synthesis is increased on treatment with some hormones such as insulin and growth hormone. They act in part by increasing the rate of elongation but the exact site of action, be it either on the ribosome or factors is not yet certain.

There has also been considerable debate as to whether all the mRNAs in a given cell are elongated at the same rate. For example it is known that reticulocytes contain 1.7 times more mRNA for α than for β globin, and yet the two proteins are made in approximately equal amounts. These findings could be explained if protein synthesis on the α mRNA was either initiated or elongated about 60% more slowly than that on β mRNA. Sucrose gradient analysis shows that α globin is made on polysomes containing on average 3 ribosomes, whereas β globin polysomes contain on average 5 ribosomes. The rate of elongation can be measured by radioactively labelling reticulocytes for very short periods and determining the incorporation of label into the various tryptic peptides of the two globins. Knowing the order of the peptides along the polypeptide chain and the time taken to label from one peptide to another an estimate of the translation time is produced.

Most recent estimates suggest that the rate of elongation of both α and β chains is approximately 50 amino acids per min. at 25°C and that the rate of elongation is constant throughout the entire length of the mRNA. Such data indicates that the rate of initiation determines the amounts of the two globins made, and subsequent experiments in cell-free systems support such a model [17].

The possibility that qualitative control of protein synthesis occurs at the elongation step in the synthesis of some proteins still remains. For example the rate of translation of a given mRNA could be dependent on the concentration of the least common amino acyl-tRNA species and this prediction leads to the idea of tRNA modulation, that is the relative amounts of different tRNA species varies depending on the particular needs of the cell. At certain stages in the development of the silkworm *Bombyx mori*, 70% of the protein made in the silk gland is fibroin and this contains 46% glycine and 29% alanine. Examination of the tRNAs and synthetases present in the gland at this time have shown that these are grossly altered and reflect the increased need for tRNA^{Gly} and tRNA^{Ala}. A similar unusual distribution of tRNA species is also found in other highly differentiated organs producing proteins of unusual composition, for example the oviduct and brain.

A similar phenomenon may also occur during phage infections. T4 codes not only for several species of virus-specific tRNA but also for an RNAse activity that specifically cleaves one host tRNA^{Leu}. In this way protein synthesis in the infected cell is turned over to production of phage proteins and host protein synthesis is inhibited. A related phenomenon may occur in animal virus-infected cells (Fig. 3.7). Some viruses contain RNA within the virion, sometimes H-bonded to the genomic RNA (RNA tumour viruses) and sometimes covalently attached to one end (picornaviruses). Whilst some of the tRNAs probably have other functions (in priming the enzyme reverse transcriptase, and in RNA synthesis) it is possible they also have a second role in elongation [8]. It is possible that the protein induced by the anti-viral agent, interferon, also affects the elongation step. In this case viral protein synthesis is specifically inhibited and one mechanism that has been proposed is that tRNA species specifically required for viral synthesis, but not for host protein synthesis, are destroyed on treatment with interferon.

Elongation during protein synthesis in animal cells occurs on two distinct populations of ribosomes, viz. the free and membrane-bound polysomes [18]. There have been many claims that there are differences in antibiotic sensitivity, and other physical properties of the two populations but it is still unclear what dictates whether any one particular ribosome will be membrane bound or not. It is equally uncertain what decides whether an mRNA will be translated on a free or membrane-bound polysome; as a general rule, however, proteins destined for export from the cell are synthesized on membrane-bound polysomes, and proteins for internal use are made on the free fraction. Thus the liver has a relatively high proportion of rough endoplasmic reticulum, which is composed of membranous structures with many ribosomes attached, and this correlates with the large amount of protein made in the organ for export into the blood. The biosynthesis of serum albumin, for example, has been studied in detail. The protein has been shown to be made exclusively on the membrane-bound polysomes, from where the nascent protein is pushed into the cisternae to be transported, processed and finally either stored or exported. Ferritin, on the other hand, is a liver protein which is required for internal use. It is made on free polysomes. The definition that proteins made on membrane-bound polysomes are exclusively for export may prove to be too rigid; many cultured cells excrete very little protein and yet they still contain a proportion of bound ribosomes.

Several suggestions have been put forward to explain the molecular mechanism which decides where a particular protein will be made. It could be that the mRNA itself has distinguishing features that cause it either to be transported directly through the endoplasmic reticulum to its site of translation, or cause it to attach to membranes after transport in the usual manner. Perhaps some unusual nucleotide sequence or protein attached to the mRNA is involved in this. An alternative explanation is that all mRNAs first initiate synthesis on free ribosomes, but that the first few amino acids polymerized, perhaps because of their hydrophobic nature, sometimes cause the nascent chain to attach to membranes, and synthesis then continues on the membrane. Such a model might explain why the N-terminal portion of many proteins is cleaved off soon after synthesis. Perhaps its only function is to provide a membrane binding site, and consequently it is not required after completion of the peptide chain. Immunoglobin light chains are an example of a protein made on membrane-bound ribosomes which has such an N-terminal region. The peptide that is cleaved off, contains a relatively high proportion of hydrophobic amino acids, particularly leucine. Several animal virus proteins also have rapidly metabolized N-terminal sequences; this may correlate with the finding that viral protein synthesis usually occurs on membranes.

3.5 Termination of protein synthesis and post-translational modification

It seems likely that the termination of protein synthesis involves a hydrolysis step in which water reacts with the growing polypeptide in very much the same way and at about the same rate as an amino acid. There does not seem to be an accumulation of completed, unreleased polypeptide nor does it seem likely that any specific control reactions occur at this stage. After chain termination the ribosomes are released from the messenger and enter a pool awaiting re-initiation. It is likely that the nearest ribosome binding site to a released ribosome is that on the mRNA from which it has just detached. It is quite probable, therefore, that any given ribosome translates the same mRNA more than once.

There has been considerable debate as to the form of ribosome that is released from the polysome after chain termination. Using ribosomal subunits labelled either with heavy or light isotope it was possible to show that subunits do exchange between rounds of synthesis and the finding that free subunits always occur in cell lysates suggested that 50S and 30S subunits are released after termination. However, the later finding that many of the experimental manipulations used cause dissociation discredited this idea. That initiation factors themselves have dissociation activity lead to the model

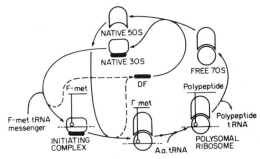

Fig. 3.8 The ribosome cycle in protein synthesis (from [19].

shown in Fig. 3.8, which suggests that 70S ribosomes are released and that these are enzymatically dissociated into subunits prior to re-initiation [19].

Once released, the nascent protein is not necessarily in its final form. Alterations can be made to the size of the primary translation product by proteolytic cleavage. The polypeptide takes up its secondary and tertiary structure and side chain modifications such as phosphorylation, glycosylation, methylation and disulphide bridges are introduced. Some post-translational modifications, such as the folding of the amino acid backbone into its three-dimensional shape, seem to occur spontaneously. Thus any given sequence of amino acids take up a structure of minimum energy, with most polar groups exposed to the aqueous environment, and most non-polar groups in the interior of the molecule where they frequently form H-bonds with one another. The formation of disulphide bridges also appears to occur mainly as a consequence of the primary sequence and hence, overall shape. A disulphide exchange enzyme randomly makes and breaks different bridge combinations until a conformation corresponding to the low energy native state is achieved. Other post-translational modifications on the other hand, such as alterations to side chains, must involve recognition of specific amino acid sequences or particular shapes within

	Glu/Asp	Thr	Ser	Ala	Met
Prokaryotes					
E.coli	4	8	20	30	48
B. subtilis	3	4	24	56	13
C. Kluyveri	6	5	22	23	44
Pseudomonas B.12	6	5	28	56	5
Eukaryotes					
S. carlsbergensis	20	6	18	50	0.3
C. vulgaris	21	3	10	66	0.3
T. pyriformis	30	6	22	32	0.3

Fig. 3.9 Amino acids found at the N-terminus of cellular proteins from different organisms (from [20]).

the proteins by the modification enzymes.

Proteolytic cleavage commonly occurs in the conversion of a nascent polypeptide to its native state. Many post-translational cleavages occur whilst the nascent protein is still ribosome-bound. For example the initiating methionine residue of eukaryotic proteins is almost always removed, often when only 15–30 amino acids have been polymerized. In bacteria the formyl group is usually removed from the nascent polypeptide, but in many cases the methionine residue remains as the N-terminal residue. This difference between prokaryotes and eukaryotes is reflected in the analysis of N-terminal residues of total protein from say *E.coli* and *Tetrahymena pyriformis,* where methionine accounts for 38% and 0.3% of all proteins respectively (Fig. 3.9). Some internal protecolytic cleavages also occur before completion of the growing polypeptide chain, additional cleavages sometimes occur soon after release, whereas others can be delayed even longer.

EMC virus provides a good example of such differences in the time of proteolysis. EMC virions contain a single species of ssRNA of molecular weight 2.5×10^6 daltons. The viral

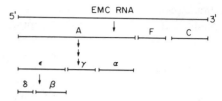

Fig. 3.10 Post-translational cleavage of nascent EMC polyprotein to give viral proteins (from [21]).

RNA is infectious and the same sense as that found on infected cell polysomes. The RNA has only one site for the initiation of protein synthesis and it codes for one large polyprotein which is subsequently cleaved to give the various stable viral proteins. Fig. 3.10 shows some of the events in polyprotein cleavage. The first step to give A occurs before the ribosome has completed translation of the entire mRNA, and the complete polyprotein (including A, F and C) can only be detected in cells treated in order to inhibit such cleavages using, say, amino acid analogues, inhibitors of proteolysis (e.g. TCPK), high temperature, or simply Zn^{++} ions. The cleavage of A to give ϵ, γ and α occurs after release and involves several steps but the final cleavage of ϵ to give γ and β only occurs after assembly of the procapsid, upon addition of the viral RNA to give a completed virion. That the latter cleavage is different is emphasized by the finding that EMC viral RNA is translated upon injection into frog oocytes and all the cleavages occur except the step $\epsilon \to \gamma$ plus β. It is not yet known if the cleavage involves recognition of a specific amino acid sequence or is dependent on the overall structure of the polyprotein.

Many other viral proteins are made in precursor form, some involving a polyprotein (such as the arboviruses and RNA tumour viruses), others precursor forms with a region at the N-terminus which is subsequently removed (e.g. reovirus and adenovirus). Yet others involve cleavage reactions at assembly (e.g. vaccinia, influenza).

The site of the bulk of the proteolytic cleavages probably involves some interaction with membranes. This may be related to the fact that viral protein synthesis often occurs on bound polysomes, that such cleavages are difficult to reproduce *in vitro* and that treatments to disrupt membranes usually inhibit cleavage. The reason that animal viruses use a mechanism involving polyprotein cleavage rather than initiating the synthesis of the various proteins at different sites on the mRNA, as in the RNA phage is unclear. At one time it seemed that there was a restriction on eukaryotic mRNAs with more than one initiation site. However, RNA phage mRNAs with multiple initiation sites are translated in animal cell-free systems and some of the animal viruses have recently been shown to have mRNAs with more than one initiation site (e.g. SFV, see Section 3.3.3 p. 51).

Many hormones and enzymes are also made in precursor form, in this case the cleavage intermediates may have specific functions in storage, transport and activation. Thus chymotrypsinogen is the inactive storage form of the proteolytic enzyme chymotrypsin.

3.6 RNA phage protein synthesis

In the earlier sections general comments have been made about each of the steps of protein synthesis and possible control at each step. This gives little idea of how each of the individual reactions are inter-related in the biosynthesis of a given protein. An example of the synthesis of a specific group of proteins which are understood in considerable detail will now be presented in the hope of giving an overview of control at the translational level. The system to be described is *E.coli* infected with phage Qβ.

On infection, viral RNA is inserted into the host cell, whereupon it is either destroyed by cellular RNAses or it binds to ribosomes and begins to make protein. The binding to ribosomes has been described in detail above (see Section 3.3.3 p. 52). To summarize, ribosomes interact with the mRNA by recognizing a common sequence near to the initiation site which forms hydrogen bonds with the $3'$-end of 16S RNA. The number of such hydrogen bonds varies because the nucleotide sequence around the initiation sites is different, but binding to coat protein sites occurs, whereas binding to both the polymerase and the A-sites is not possible because they are made inaccessible by secondary structure. The mRNA:rRNA interaction is stabilized by initiation factors (IF.3), which may or may not be different for each site, and an initiation complex formed. The rate of initiation complex formation dictates the amount of each of the viral proteins and this rate is dependent on both effects mentioned above. Soon after initiation from the coat protein site, the secondary structure of the mRNA is disrupted by the passage of the ribosome along it. The polymerase initiation site becomes exposed and ribosomes attach to it, again at a rate dependent on the interaction with both rRNA and IF-3.

The RNA polymerase next begins to replicate the viral RNA (Fig. 3.11). Here a possible problem arises, particularly early in infection when the number of RNA molecules is low, because RNA polymerase synthesizes negative strand RNA in the direction $5' \rightarrow 3'$, beginning from the $3'$-end of the input plus strand RNA. In doing so it risks meeting a ribosome travelling in the opposite direction, translating the RNA. Such a situation does not in fact arise, because the RNA polymerase first binds to a site on the RNA, not as might have been expected near to the $3'$-end, but near the coat protein ribosome binding site. This effectively blocks further initiation at this site and once all the ribosomes have been cleared from the RNA the polymerase begins

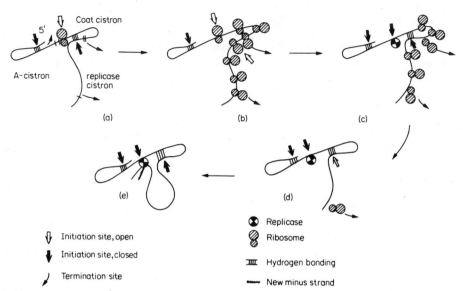

Fig. 3.11 Translational control in the synthesis of Qβ proteins (from [22]).

RNA synthesis, with no danger of interruption. It is interesting that active Qβ polymerase consists of four subunits, only one of which is virus-coded. Surprisingly, the three others are proteins known to function in protein synthesis, Tu and Ts and the ribosomal protein S1. Their function in RNA synthesis is not certain, but their role in protein synthesis is well established. S1 for example is bound near the 3′ end of 16S rRNA where it influences the interaction between rRNA and initiation site sequences. Perhaps S1 has a role in binding polymerase to mRNA, similar to its role in binding ribosomes to mRNA. The presence of molecules required for protein synthesis in RNA polymerases emphasizes the interdependence of the two processes. Possibly the relative amounts involved in protein and RNA synthesis at any one time is a major influence in controlling cellular metabolism.

Later in infection, when large amounts of RNA have been produced, there is a need for large amounts of coat protein to form progeny virions, but relatively little need for more polymerase; in other words the coat protein is structural and therefore required in stoichiometric amounts, whereas polymerase is an enzyme and required in only catalytic amounts. This is to some extent allowed for in the different rates of initiation at the two sites. However a further control mechanism also operates to reduce polymerase synthesis and results from the binding of coat protein to the mRNA at a site near to the polymerase initiation site. Bound coat protein inhibits initiation at that site, and as a result the amount of polymerase made late in infection is reduced.

Thus by this very subtle combination of controlling elements the viral proteins are synthesized in the ratio of about 20: 5: 1, coat: polymerase: A-protein, which presumably is the optimal ratio for successful infection. It should be emphasized that the effectors used involve negative as well as positive control, in switching the available ribosomes to the appropriate initiation site at the correct time. Control in this case is entirely self-regulating.

Few other systems are understood in such detail, but our knowledge of protein synthesis is now sufficiently advanced that we can conceive equally plausible mechanisms to control protein synthesis, in which control is also influenced by external signals, for example during hormonal regulation, during early development and many other physiological states. Each would presumably involve subtle changes in the activity of the various known components of protein synthesis. In the future, it will be interesting to learn if similar mechanisms do occur in these more complex situations, or if Nature has yet more surprises in store for us.

References

[1] Darnbrough, C., Legon, S., Hunt, T. and Jackson, R.J. (1973), *J. Mol. Biol.,* **76**, 379-403.
[2] Revel, M. (1972), in *Mechanism of Protein Biosynthesis and its Regulation,* Bosch, L., (ed.), North-Holland, Amsterdam, pp. 87.
[3] Bretscher, M.S. (1968), *Nature,* London, **218**, 675-677.
[4] Legon, S., Jackson, R.J. and Hunt, T. (1973), *Nature New Biol.,* **241**, 150.
[5] Ahern, T., Sampson, J. and Kay, J.E. (1974), *Nature,* **248**, 519-521.
[6] Dintzis, H.M. (1961), *Proc. Nat. Acad. Sci. (USA),* **47**, 247-261.
[7] Brown, J.C. and Smith, A.E. (1970), *Nature,* **226**, 610-612.
[8] Smith, A.E. (1975), in 25th Symposium of the Society of General Microbiology, *Control Processes in Virus Multiplication,* Burke, D.C. and Russell, W., (ed.), Cambridge University Press, pp. 183-223.
[9] Lodish, H.F. (1970), *Nature,* London, **226**, 705-707.
[10] Shine, J. and Dalgarno, L. (1975), *Nature,* London, **254**, 34-38.
[11] Revel, M. (1972), *FEBS Symposium,* Vol. **23**, p. 237, Academic Press, London.
[12] Wigle, D.T. and Smith, A.E. (1973), *Nature New Biol.,* **242**, 136-140.
[13] Heywood, S.M. (1970), *Proc. Nat. Acad. Sci. (USA),* **67**, 1782-1788.
[14] Nakajima, K., Ishitsuka, H. and Oda, K. (1974), *Nature,* London, **252**, 649-653.
[15] Heywood, S.M., Kennedy, D.S. and Bester, A.J. (1974), *Proc. Nat. Acad. Sci., (USA),* **72**, 2428-2431.
[16] Lodish, H.F. (1974), *Nature,* London, **251**, 385-389.
[17] Lodish, H.F. (1971), *J. Biol. Chem.,* **246**, 7131.
[18] Sabatini, D., Borgese, N., Adelman, M., Kreibich, G. and Blobel, G. (1972), in *FEBS Symposium,* Vol. 27, North-Holland, Amsterdam, pp. 147.
[19] Subramanian, A.R., Ron, E.Z. and Davis, B.D. (1968), *Proc. Nat. Acad. Sci. (USA),* **61**, 761.
[20] Sarimo, S.S. and Pine. M.J. (1969), *J. Bacteriol.* **98**, 368-374.
[21] Butterworth, B.E. and Reuckert, R.R. (1972), *J. Viro.* **9**, 823-828.
[22] Weissmann, C., Billeter, A.M., Goodman, H.M., Hindley, J. and Weber, H. (1972), in *FEBS Symposium,* **27**, North-Holland, Amsterdam, pp. 1.

Index

Adaptor hypothesis, 29
Aminoacyl site (A-site), 19, 21, 36, 41, 43, 49
Aminoacyl tRNA synthetases, 33, 35, 38
Antibiotics,
   chloramphenicol, 38
   cycloheximide, 38
   edeine, 47
   erythromycin, 27
   fusidic acid, 28, 43
   kasugamycin, 27
   puromycin, 20, 27, 49
   spiramycin, 27
   streptomycin, 27
   thiostrepton, 43

Bacteriophage,
   T4, 10, 24, 58
   Small RNA, 10, 14, 37, 54
Bacteriophage RNA,
   cistron order of, 15
   coat protein gene sequences, 16, 50
   control of translation of, 53, 54, 62
   ribosome binding sites of, 14, 52, 54, 62

Cell-free systems, 10, 18
Colicin E3, 53

Elongation factors,
   assay of, 20, 41
   comparison with initiation factors, 43
   interaction with initiator tRNA, 38, 42
   interaction with tRNA, 30
   properties of factors G and EF2, 42
   properties of factors T, Ts, Tu and EF1, 41
   role in control of translation, 57
   role in transcription, 62
Elongation of polypeptide chain
   control of rate, 57
   overall reaction scheme, 48

Fragment reaction, 21, 43

Genetic code, 16, 20, 29
GMP.PCP, 50
GTP
   number hydrolysed per peptide bond, 42
   role in binding tRNA, 27, 41
   role in initiation, 28, 43, 49
   role in termination, 28, 44
   role in translocation, 28, 42

Hn RNA, 17, 47
Hormones, 46, 57

Initiation codon AUG
   in bacteriophage mRNA, 14
   role in phasing mRNA, 37, 51
Initiation complex
   effect of RNA polymerase and coat protein, 51, 61
   order of addition of components, 47
   ribosome binding sites from, 14, 52
   role of initiation factors in formation of, 49, 53
   role of initiator tRNA in formation of, 47
   role of mRNA in formation of, 50
   role of rRNA in formation of, 52
Initiation factors
   Factor F_1, 43
   Factor F_2, 43, 49
   Factor F_3, 43, 53
Initiator tRNA
   primary sequence of, 39
   properties of, 37
   role in initiation complex formation, 14, 20, 32, 47, 55
Interferon, 58

Methods
   column chromatography, 13, 24, 31, 41
   electron microscopy, 9, 17
   NMR, 36
   polyacrylamide gels, 12
   protein fingerprinting, 13
   RNA sequencing, 14, 16, 31
   X-ray crystallography, 34, 41
Modified nucleotides
   in mRNA, 17, 51
   in rRNA, 23, 27
   in tRNA, 29, 31, 37
mRNA
   blocked ends, 17, 51
   polyA, 17, 51
   properties of, 9, 50
   proteins bound to, 10, 16, 47
   ribosome binding site of, 14, 52
   structure of, 13

Nonsense mutations, 37

Peptidyl transferase, 19, 43
Peptidyl site P-site, 19, 21, 36, 41, 48
PolyA, 10, 17, 47, 51
Polysomes
   electron microscopy of, 9
   membrane bound, 58
   preparation of, 18
Post-translational modification, 59

Ribosomal proteins
   chemical modification of, 25
   cross linking of, 25, 54
   properties of, 22, 24
   proteins L7/112, 27, 43
   role in control of translation, 53
Ribosomal RNA
   5S RNA
      primary sequence, 22
      role in tRNA binding, 22, 32, 35
   16S RNA
      role in binding tRNA, 36
      role of 3'end in mRNA binding, 15, 24, 52
      sequences in, 23
   23S RNA, 23
Ribosomal subunits
   preparation of, 18
   role in initiation, 43, 47
   subunit cycle, 59
Ribosomes
   assembly map, 26
   biosynthesis, 28
   partial reactions of, 19
   properties of, 17
   reconstitution of, 25
   role in initiation complex formation, 53

Suppression, 37

Termination factors, 44
Termination of polypeptide synthesis, 60
Transfer RNA
   classification of, 33
   cross linking, 27, 36, 40
   modulation of, 57
   mutants of, 36
   primary sequence of, 30
   properties of, 20, 29
   purification of, 31
   tertiary structure of, 33, 35
Transformylase, 37
Translational control RNA, 55
Translocation, 21, 28

Viruses
   adenoviruses, 55, 60
   picornaviruses, 10, 17, 55, 58, 60
   poxviruses, 61
   reoviruses, 10, 17, 60
   RNA tumour viruses, 58, 60
   Semliki forest virus, 51, 60
   SV40, 55

Wobble hypothesis, 29

Xenopus oocytes, 11, 17, 22, 56, 60

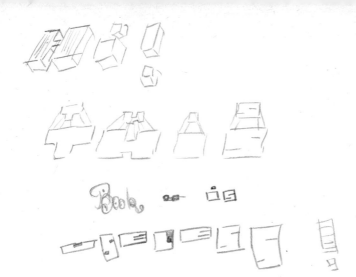

Don't Read It